Visit our website

to find out about other books from Churchill Livingstone and our sister companies in Harcourt Health Sciences

Register free at
www.harcourt-international.com

and you will get

- the latest information on new books, journals and electronic products in your chosen subject areas
- the choice of e-mail or post alerts or both, when there are any new books in your chosen areas
- news of special offers and promotions
- information about products from all Harcourt Health Sciences companies including Baillière Tindall, Churchill Livingstone, Mosby and W. B. Saunders

You will also find an easily searchable catalogue, online ordering, information on our extensive list of journals...and much more!

Visit the Harcourt Health Sciences website today!

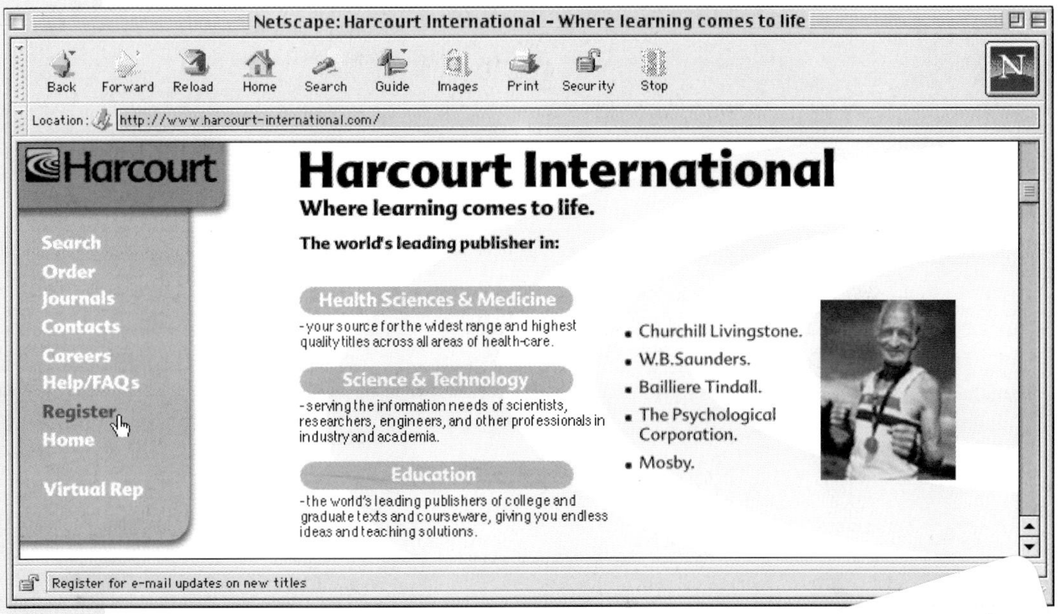

Harcourt Health Sciences

Atlas of

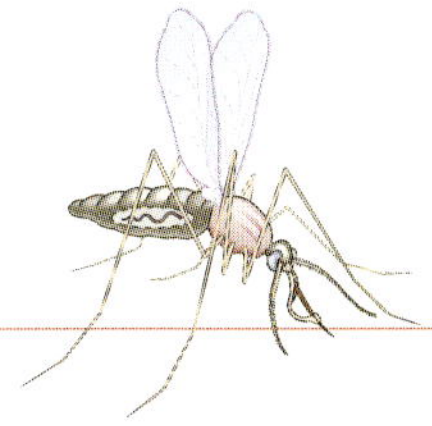

Medical Helminthology and Protozoology

Commissioning Editor: Timothy Horne
Project Development Manager: Jim Killgore
Designer: Sarah Russell
Project Manager: Nancy Arnott

Fourth Edition

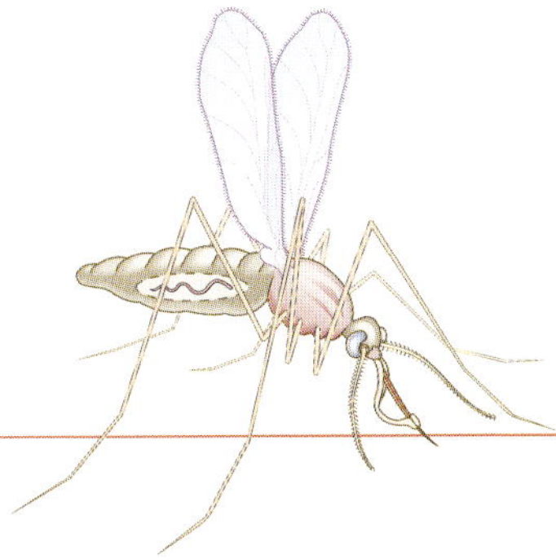

Atlas of Medical Helminthology and Protozoology

Peter L. Chiodini BSc PhD FRCP FRCPath
Consultant Parasitologist, Department of Clinical Parasitology, The Hospital for Tropical Diseases, Honorary Senior Lecturer, The London School of Hygiene and Tropical Medicine, London, UK

Anthony H. Moody MPhil MIBiol C.Biol FIBMS
Laboratory Manager, Department of Clinical Parasitology, The Hospital for Tropical Diseases, London, UK

David W. Manser BSc MSc FIBMS
Biomedical Scientist, Department of Clinical Parasitology, The Hospital for Tropical Diseases, London, UK

Illustrated by Robert Britton

CHURCHILL LIVINGSTONE

EDINBURGH LONDON NEW YORK PHILADELPHIA ST LOUIS SYDNEY TORONTO 2001

CHURCHILL LIVINGSTONE
An imprint of Harcourt Publishers Limited

First published 2001

ISBN 0443 06268 4
International Student Edition ISBN 0443 06267 6

British Library Cataloguing in Publication Data
A catalogue record for this book is available from the British Library

Library of Congress Cataloging in Publication Data
A catalog record for this book is available from the Library of Congress

Note
Medical knowledge is constantly changing. As new information becomes available, changes in treatment, procedures, equipment and the use of drugs become necessary. The authors and the publishers have, as far as it is possible, taken care to ensure that the information given in this text is accurate and up to date. However, readers are strongly advised to confirm that the information, especially with regard to drug usage, complies with the latest legislation and standards of practice.

Printed in China

The publisher's policy is to use **paper manufactured from sustainable forests**

Preface

Since this atlas was first published, major advances in immunology and molecular biology have transformed our understanding of the parasitic diseases which affect humans. The programme to eradicate Guinea worm is well advanced and real progress is being made towards a malaria vaccine. However, none of the parasites described in the first edition have yet been consigned to history. Indeed, *Cyclospora* and the microsporidia are newly recognised as important human pathogens even since the third edition, and in some geographical areas the malaria situation is worse, with the spread of multi-drug resistant *Plasmodium falciparum* malaria. There is a great deal left to be done.

Effective action against parasitic disease requires a team approach, including epidemiologists, biologists, diagnostic laboratory workers and clinicians. Common to all these disciplines is a need to understand the life cycles and morphology of the organisms they confront. It is hoped that this edition of the atlas will provide an appropriate introduction. The strong emphasis on diagnosis has been retained and since diagnostic parasitology still relies heavily on morphology, we have strengthened this area with the introduction of colour illustrations and photomicrographs.

We hope this book will help to kindle enthusiasm for the effort to control these parasites and the diseases they cause.

London
2001

P. L. C.
A. H. M.
D.W. M.

Contents

Introduction

The protozoon and the helminth, as regards tropical pathology, are in the ascendant.

Sir Patrick Manson (1899)

Parasites to the Ancient Greeks were those who sat at another's table and paid for their meal with flattery. In biology, a parasite is an animal or plant living in or on another (the host) and drawing nourishment from it. This definition could include viruses, bacteria and fungi as well as protozoa and helminths, but historically the first group has been studied in microbiology, the second in parasitology. In tropical diseases, Manson's dictum remains valid today.

Protozoa are small, unicellular organisms, which contain a nucleus and functional organelles. They reproduce quickly and asexually in the host, but may have a sexual phase of their life cycle in another host or vector.

Helminths (worms), which are metazoa, are larger, multicellular organisms, normally visable to the naked eye in their adult form. They reproduce sexually, usually within the host, and have pre-adult stages (ova, larvae) which live externally or in the host.

Transmission of parasites requires:

- a source or reservoir which may be human or animal
- a route of infection, e.g. ingestion, penetration or an insect vector.

The **definitive** host is that in which sexual reproduction occurs (e.g. humans for African trypanosomiasis). An **intermediate** host is another animal essential to the completion of the life cycle (e.g. snails for schistosomiasis).

Parasites cause disease in humans by:

- mechanical effects, e.g. hydatid cyst
- invasion and destruction of host cells, as in malaria
- allergic or inflammatory immune reaction by the host to the parasite, e.g. toxocariasis and trypanosomiasis
- competition for specific nutrients, e.g. *Diphyllobothrium latum* for vitamin B_{12}
- or there may be no obvious disease, as in *Taenia saginata* in humans.

Diagnosis in parasitic diseases depends on:

- a history of exposure and the clinical pattern of illness in the patient
- identification of the parasite itself in excreta (stool, urine), blood, or specific tissues
- indirect evidence of the parasite by testing the patient's blood for antibodies
- detection of parasite antigens in clinical specimens
- detection of parasite DNA or RNA in clinical specimens.

Helminthology

Worms of medical importance

Nematodes
(round worms)

- Unsegmented
- Possess mouth, oesophagus and anus
  - Important in further diagnosis
- In general, sexes separate
- Reproduction
  - Oviparous
  - Larviparous
- Infection by
  - Ingestion of eggs, or
  - Penetration of larvae through surfaces, or
  - Arthropod vector, or
  - Ingestion of encysted larvae

Cestodes
(tape worms)

- Segmented
- Possess scolex, neck and proglottids
- Hermaphroditic
- Reproduction
  - Oviparous
  - Sometimes multiplication within larval forms
- Infection generally by encysted larvae

Trematodes
(flukes)

- Unsegmented
- Leaf-like or cylindrical
- Generally hermaphroditic
- Reproduction (digenetic)
  - Oviparous
  - Multiplication within larval forms
- Infection mainly by larval stages entering intestinal tract, sometimes through skin

Nematode (round) worms

Enterobius vermicularis (thread or pin worm)

Life cycle

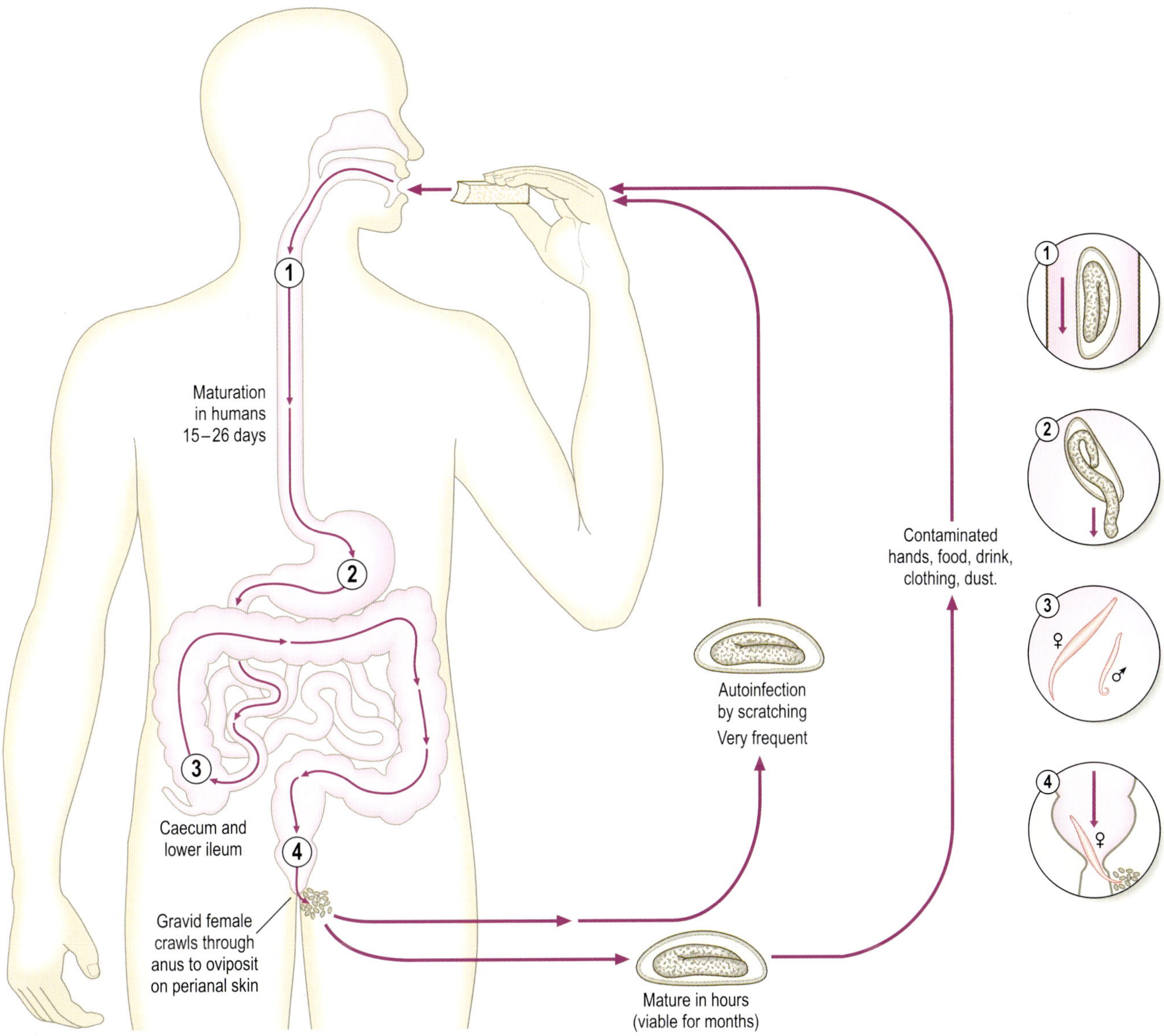

Distribution

350 million infected worldwide, often group or institutional infection.

Pathology and Clinical features

Most infections are asymptomatic. Perianal itching may be troublesome. In females, migrating worms may cause pruritis vulvae or vaginitis. Rarely, urinary tract infection or appendicitis can occur. Migration into the peritoneal cavity has been recorded.

Laboratory diagnosis

Mild eosinophilia.

Ova can be recovered from the perianal area using clear adhesive tape or a cotton swab moistened with saline. Early morning collection before washing gives best recovery. In females, ova may occasionally be recovered from urine.

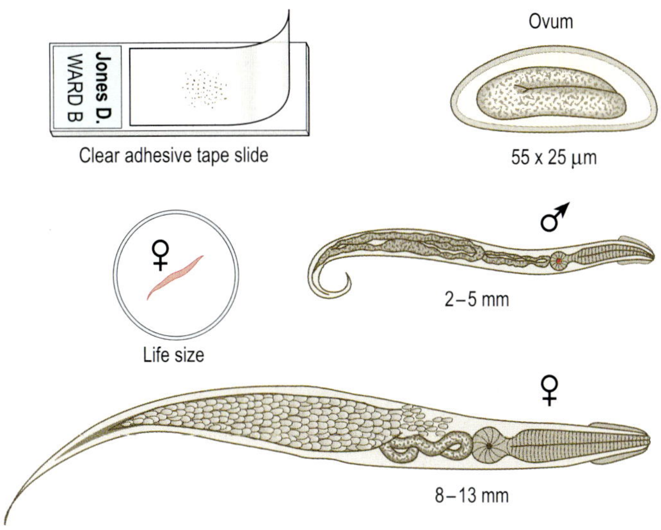

Trichuris trichiura (whip worm)

Life cycle

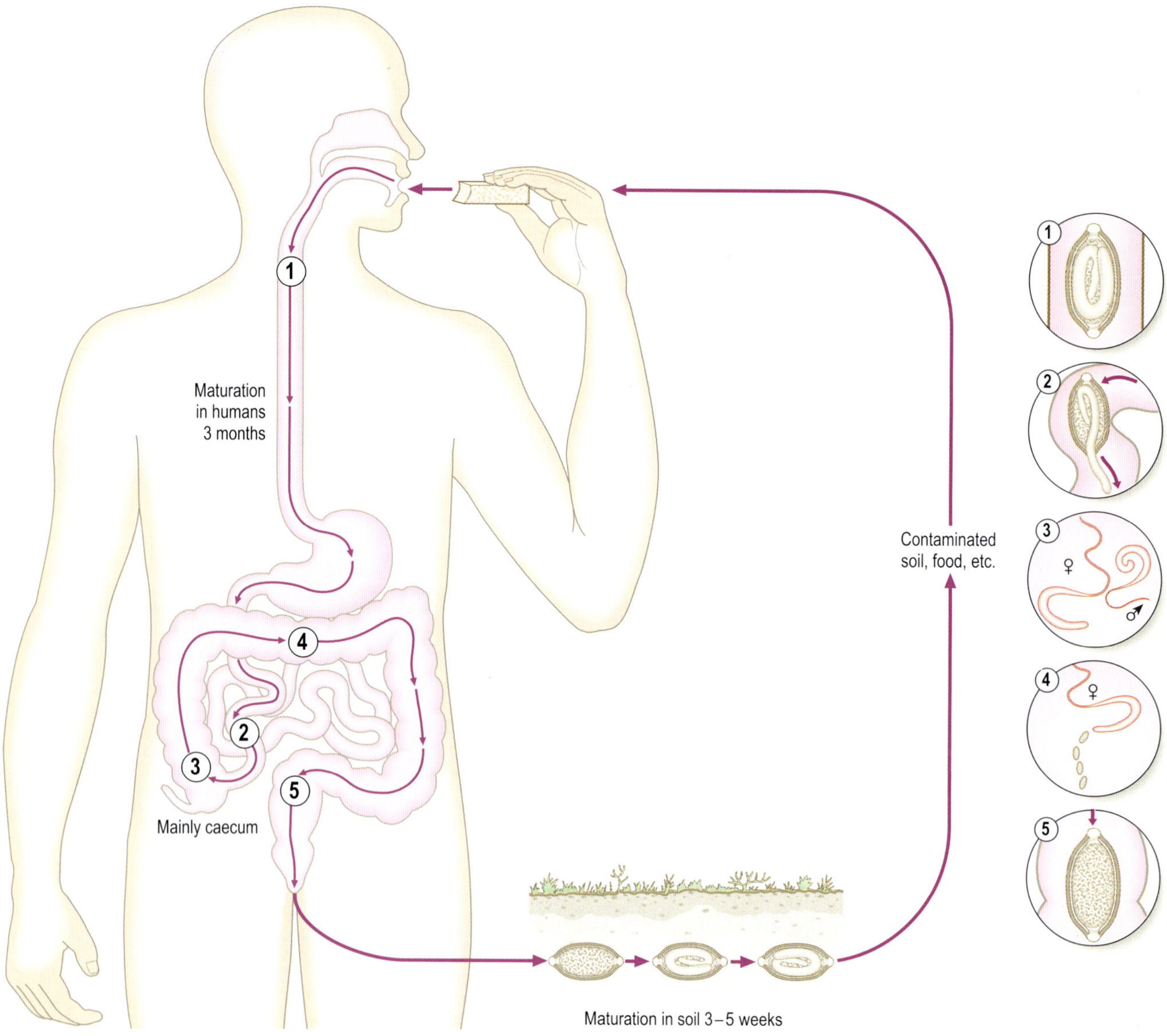

Pathology and Clinical features

Light infections may be asymptomatic. Heavy infections can result in the trichuris dysentery syndrome, rectal prolapse, rectal bleeding, anaemia, growth stunting and growth retardation in children.

Laboratory diagnosis

Eosinophilia may occur.

Ova may be recovered in faeces by concentration methods.

Distribution

1.3 billion infected worldwide.

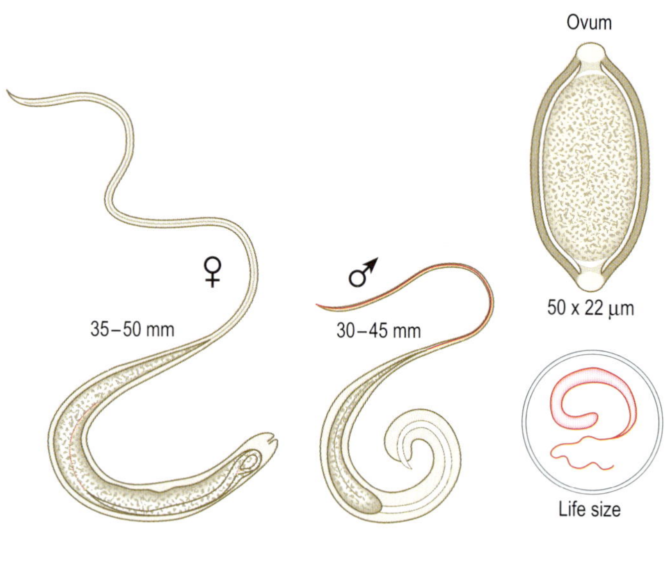

Ascaris lumbricoides (round worm)

Life cycle

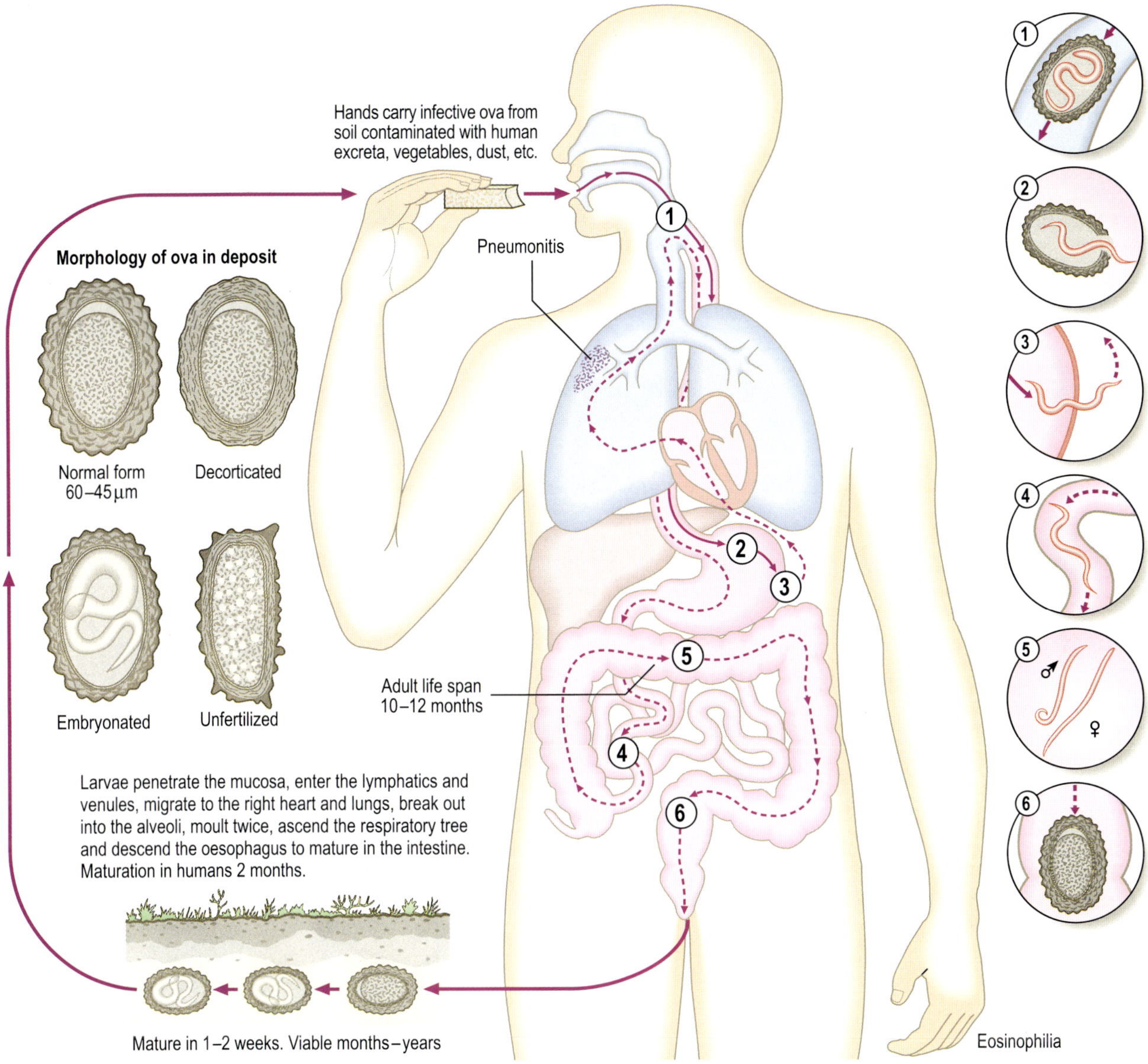

Pathology and Clinical features

Larvae can cause pneumonitis with eosinophilia. Adult worms can cause obstruction of the small intestine, bile ducts and trachea; also appendicitis, pancreatitis and peritonitis. Children may vomit up a bolus of adult worms, or cough up immature worms.

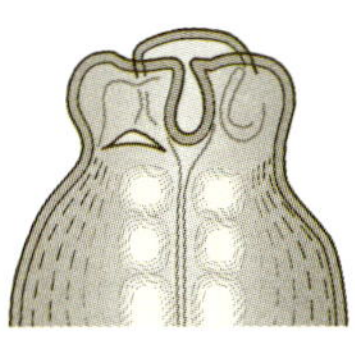

Head of adult to show arrangement of the three lips

Laboratory diagnosis

Ova may be recovered from faeces by concentration methods. Rarely larvae can be found in sputum, and must be distinguished from those of *Strongyloides*. Eosinophilia is present in the larval invasion stage.

No specific serology is currently available.

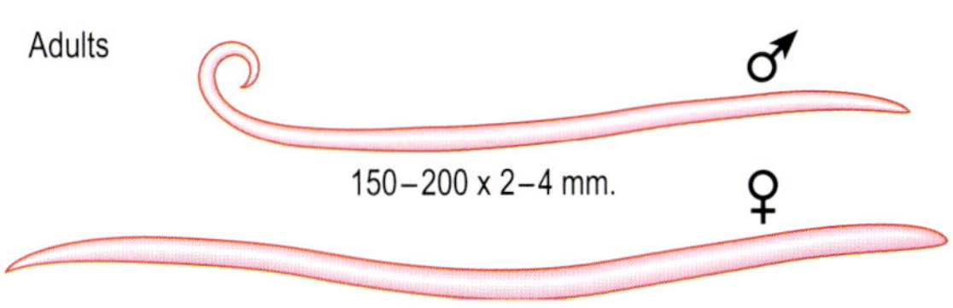

Distribution

1.47 billion infected worldwide.

Hookworms

Ancylostoma duodenale *Necator americanus*

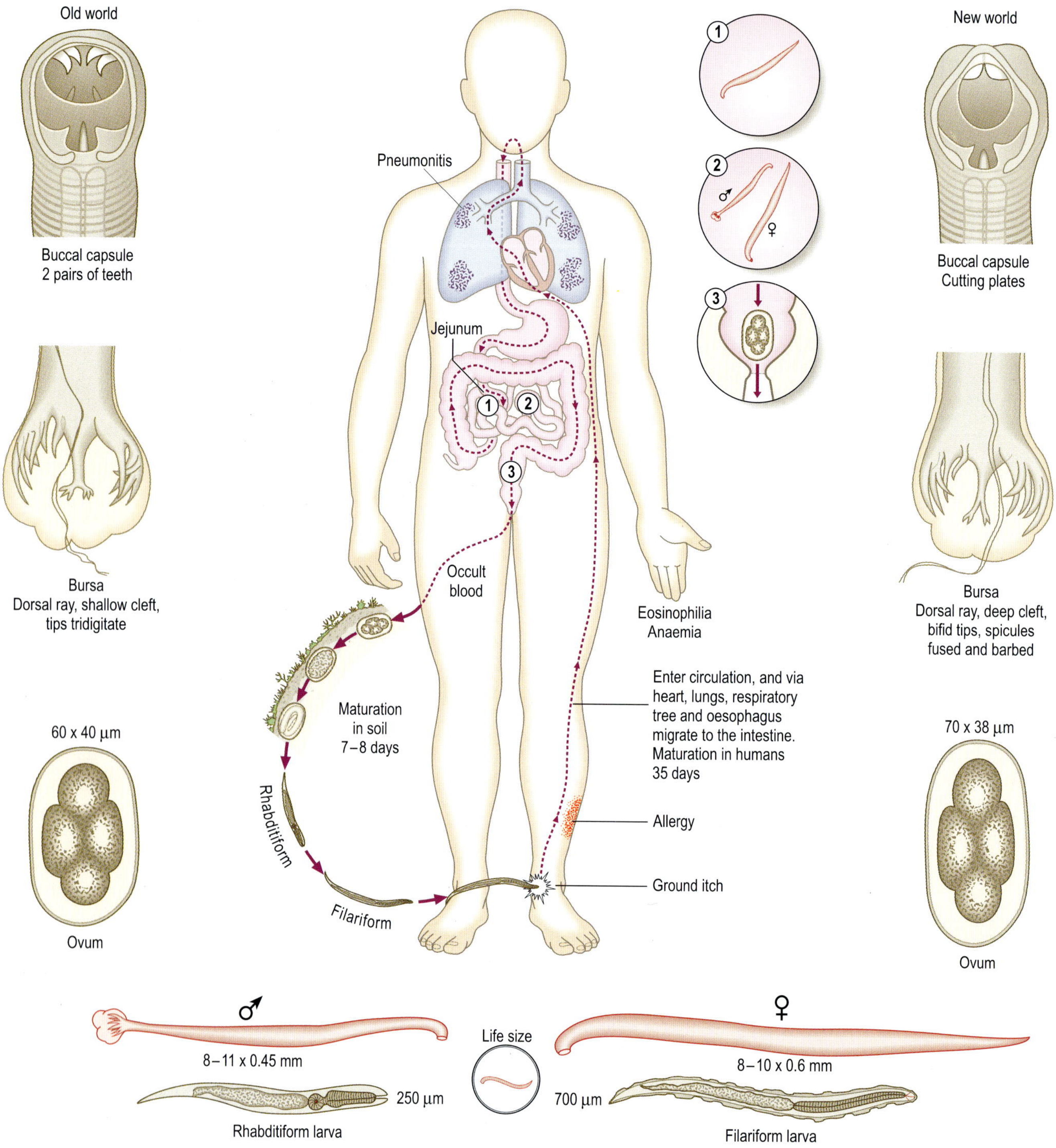

Pathology and Clinical features

Ground itch may follow skin penetration by filariform larvae. Pneumonitis can result from larval migration through the lungs. Adult worms in the jejunum ingest blood. Occult gastrointestinal bleeding occurs. Iron deficiency anaemia and its sequelae in heavy infections.

Distribution

900 million infected worldwide.

Laboratory diagnosis

Eosinophilia.

Ova may be recovered from faeces by concentration methods. Rhabditiform larvae may be seen in old faecal specimens and must be distinguished from *Strongyloides* by the appearance of the buccal cavity.

Strongyloides stercoralis

Life cycle

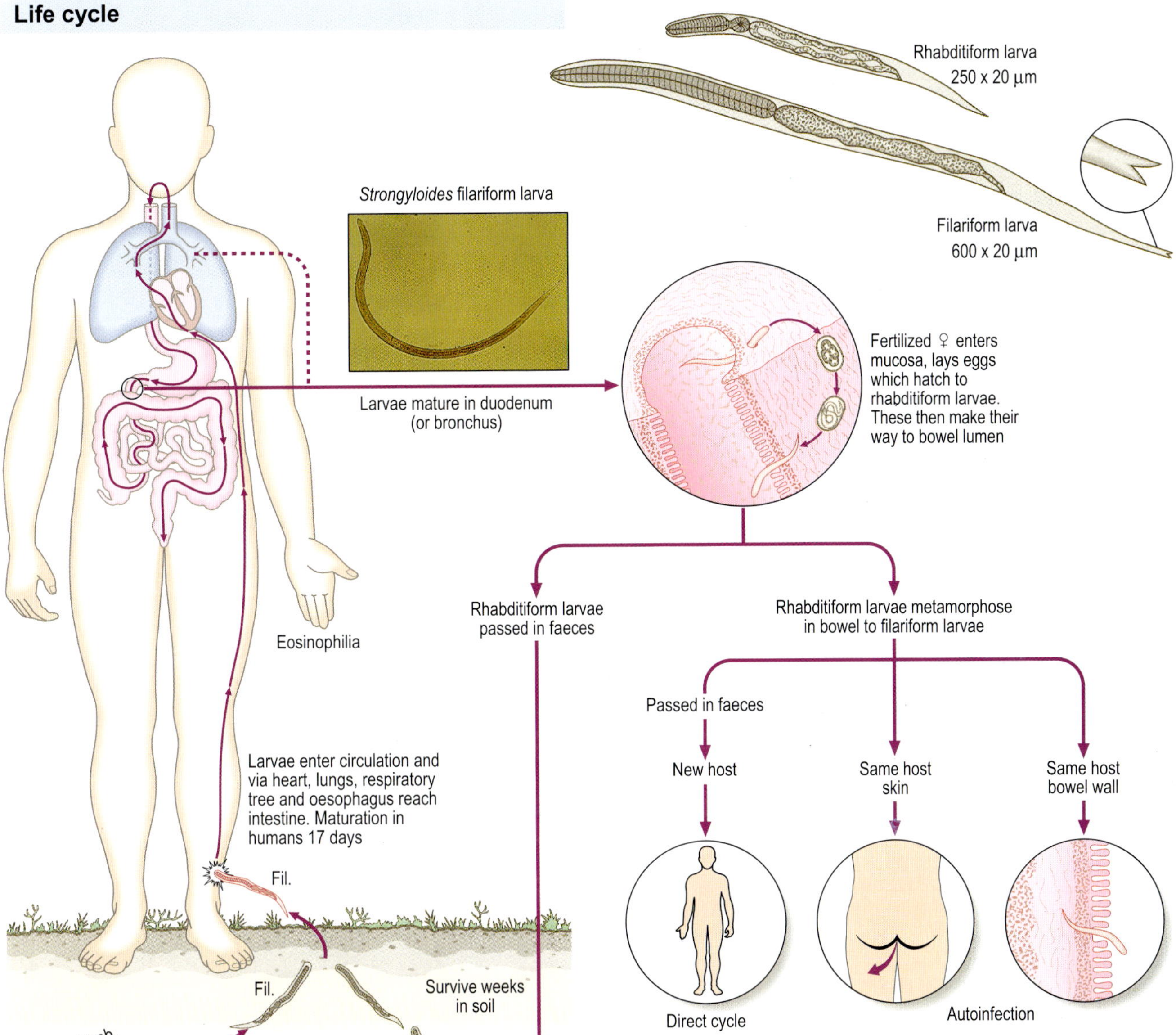

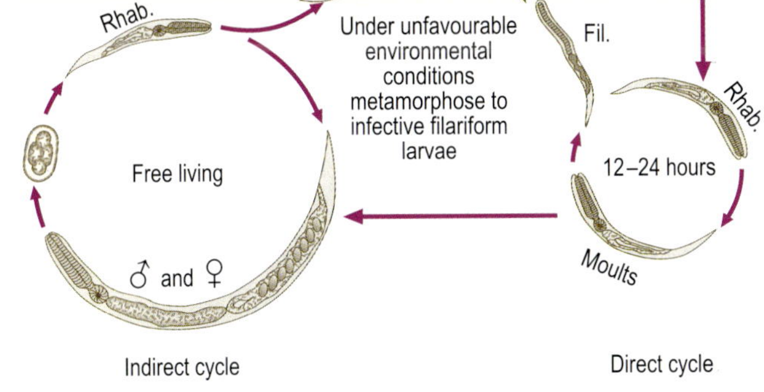

Pathology and Clinical features

Skin penetration by larvae may cause local irritation. Migrating larvae can cause pneumonitis, and ectopic larvae can sometimes be found in the brain and other viscera. A characteristic serpiginous urticarial rash (larva currens) is seen on the trunk and buttocks.

Established infection may have no signs or symptoms, or present with larva currens alone. Diarrhoea, abdominal pain, bloating and sometimes malabsorption can be found. The *Strongyloides* hyperinfection syndrome results from massive autoinfection with filariform larvae in the presence of severe immunosuppression or cachexia. Risk factors include steroid and/or cytotoxic therapy, HTLV1 infections, HIV infections, malignancy, severe malnutrition and other severe systemic disorders. Clinical features include diarrhoea, gastrointestinal haemorrhage or perforation, pneumonitis, Gram-negative bacterial meningitis or septicaemia with high mortality.

Laboratory diagnosis

Eosinophilia may be present, but its absence does not exclude diagnosis. It is essential to examine fresh specimens. Rhabditiform larvae can be seen in faeces by direct microscopy or by concentration methods. Filariform larvae may also be seen in faeces, sputum and other body fluids, particularly in immunocompromised hosts. Faecal culture using charcoal is an important diagnostic method. Duodenal aspiration and the 'string test' are also recommended isolation methods. Serology by ELISA is useful in chronic infection.

Distribution

70 million infected worldwide.

Trichinella spiralis

Life cycle

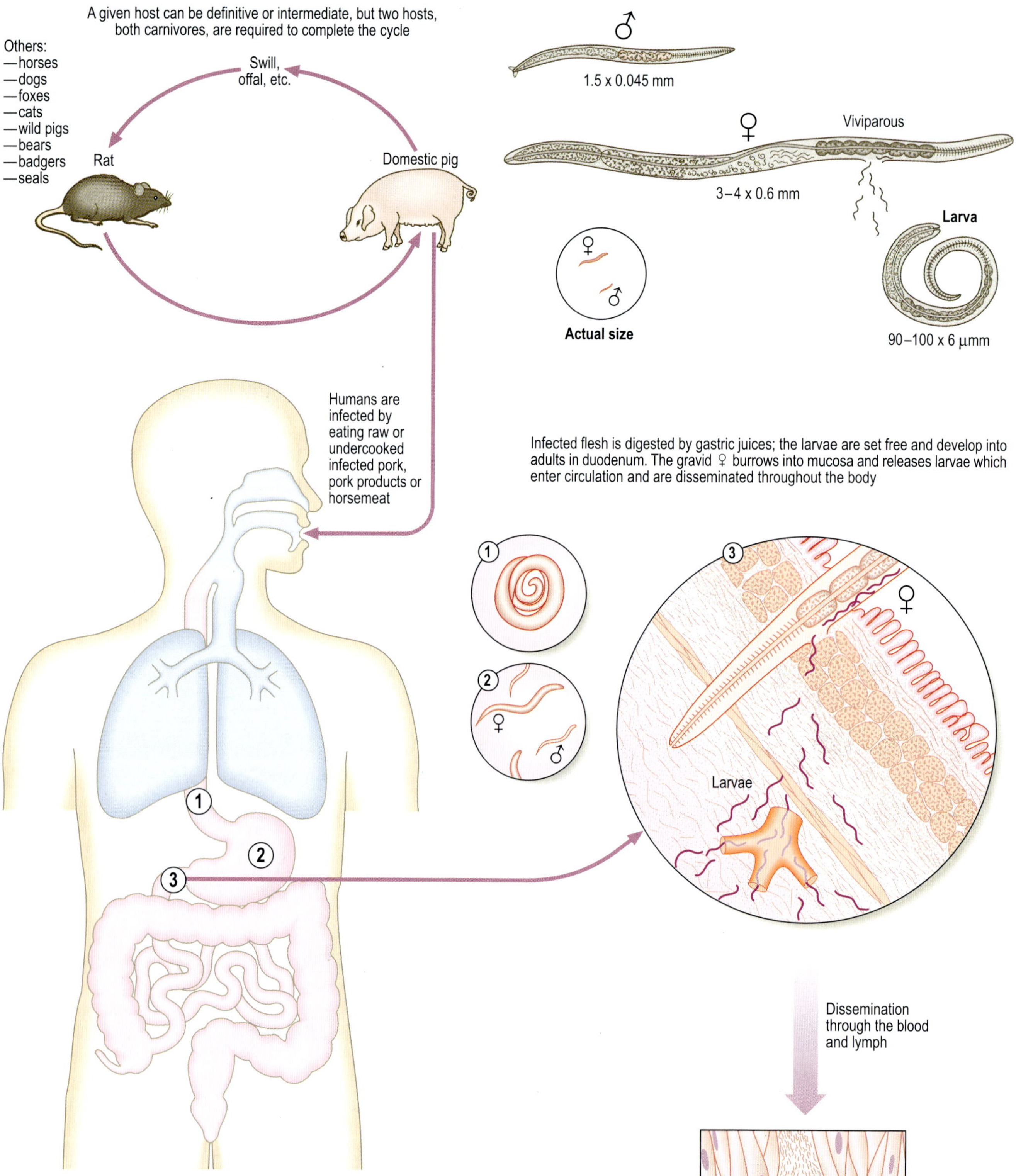

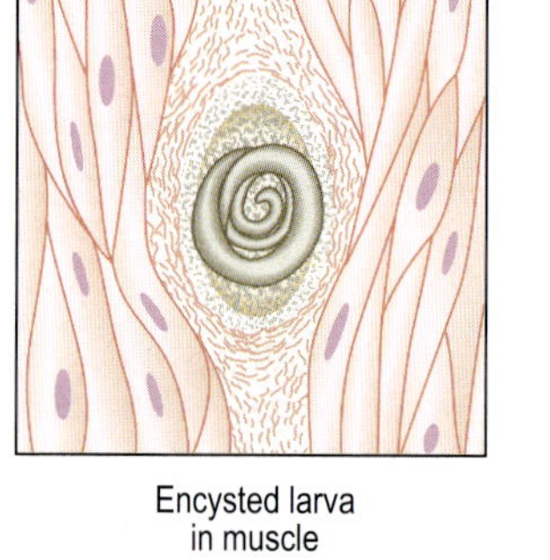

Laboratory diagnosis

Eosinophilia and high serum CPK in the acute phase. At the encystment stage, use muscle biopsy, muscle crush preparation and serology (IFAT or ELISA).

Distribution

50 million infected worldwide.

Pathology and Clinical features

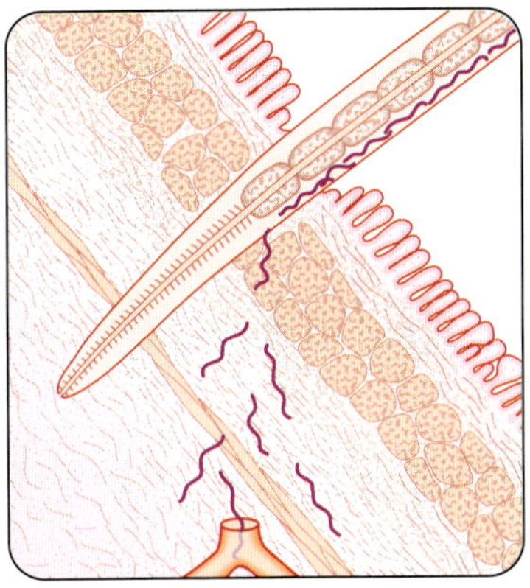

Invasion

Intestinal inflammation leading to diarrhoea. Inflammatory response leading to periorbital oedema, haemorrhages under nails, muscle pains and myocarditis.

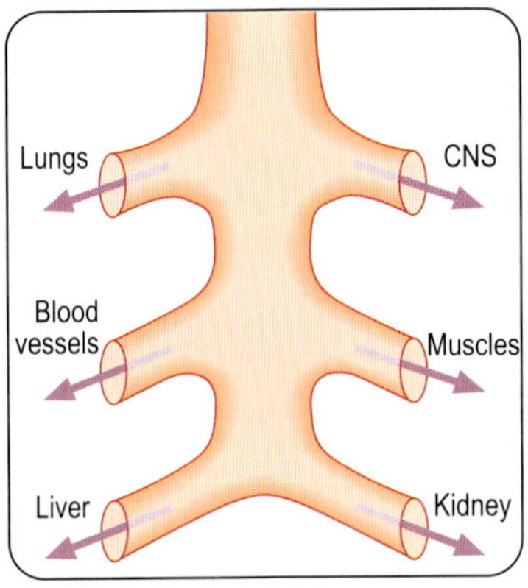

Dissemination

Migration may occur through any tissue but larval encystment is only in striated muscle. A granulomatous response develops elsewhere.

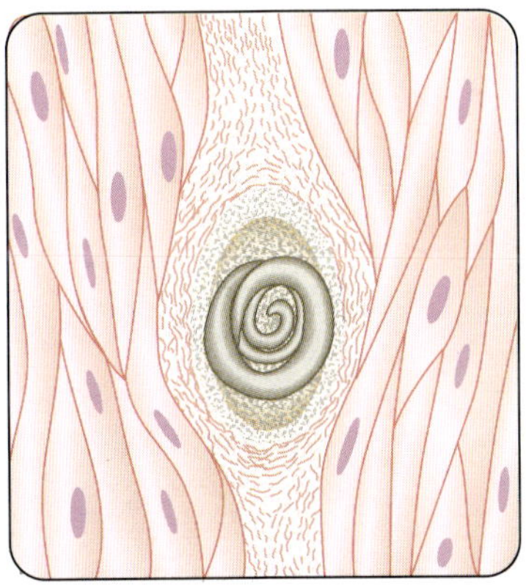

Localization

Especially muscles of respiration and tongue. Long term: eventual fibrosis and degeneration, resulting in calcification.

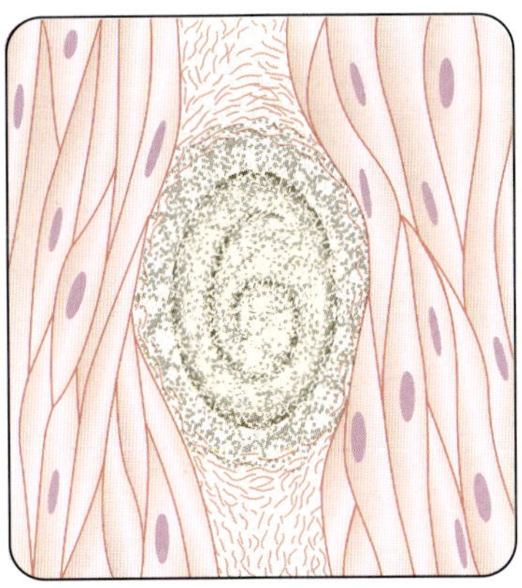

Organization

Laboratory diagnosis

At the diarrhoeal stage, adults and larvae may be found occasionally in faeces. Eosinophilia is high. At the encystment stage, use muscle biopsy, muscle crush preparation and serology (IFAT or ELISA).

Wuchereria bancrofti (filariasis)

Life cycle

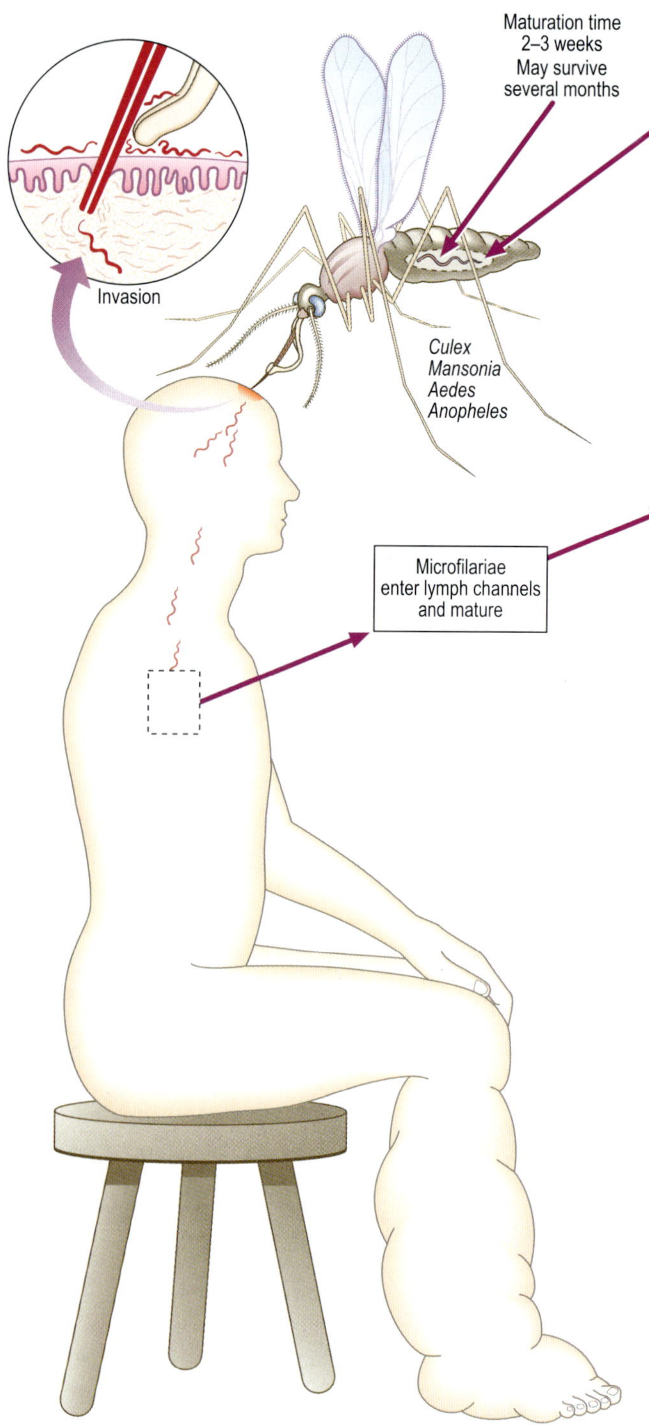

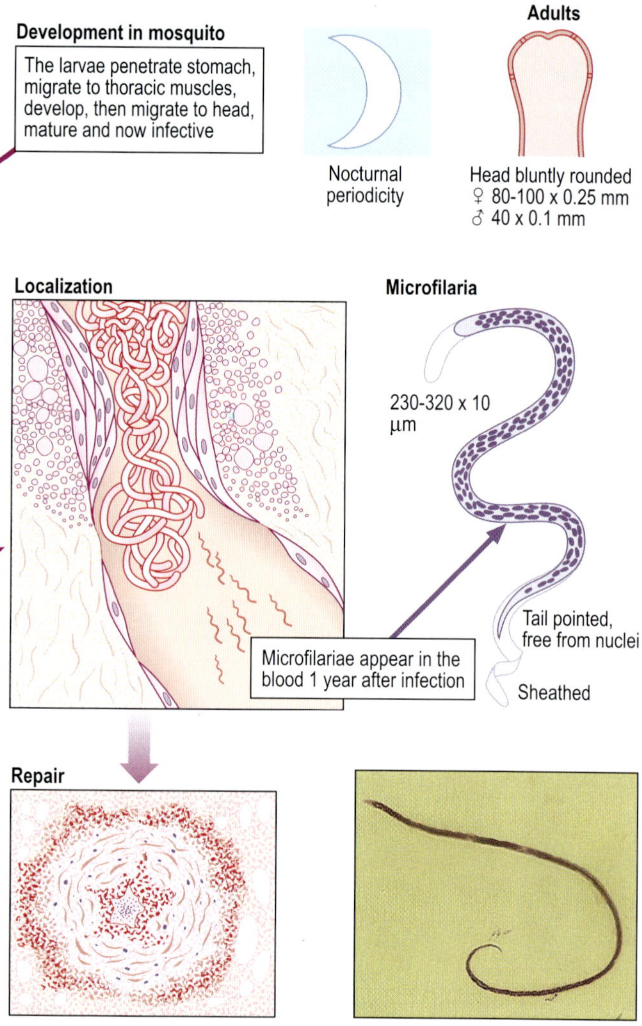

Distribution

90 million infected worldwide.

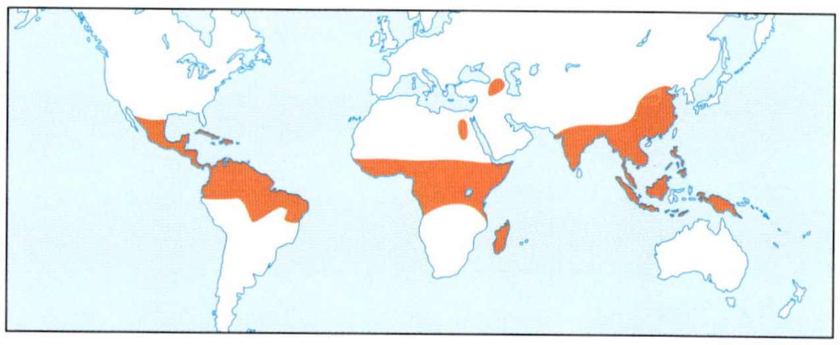

Pathology and Clinical features

Adult worms in the lymphatic channels cause proliferation of the lining of the endothelium. Surrounding infiltration of eosinophils, macrophages, lymphocytes and giant cells causes filarial granulation tissue leading to obstruction, secondary infection, fibrosis and calcification. The results of this are acute lymphangitis, filarial abscess, lymphadenopathy, elephantiasis, hydrocoele and chyluria. Tropical pulmonary eosinophilia (TPE) occurs in individuals who are hyper-responsive to filarial antigens, giving rise to nocturnal cough, wheeze and low-grade fever.

Laboratory diagnosis

Eosinophilia.

Microfilariae are found in peripheral blood collected between 10pm and 2am, or at midday for *W. bancrofti var.pacifica*. Thick blood films are examined stained or unstained, concentration by Knott's method will increase sensitivity. Filtration of citrated blood through a 5 micron pore size polycarbonate membrane is the method of choice.

Microfilariae can also be found in chylous exudate, chylous urine and in hydrocoele fluid.

Serology. ELISA is of use. Patients with TPE have high filarial antibody levels. A specific *W. bancrofti* antigen immunochromatographic test is now commercially available.

Brugia malayi

Life cycle

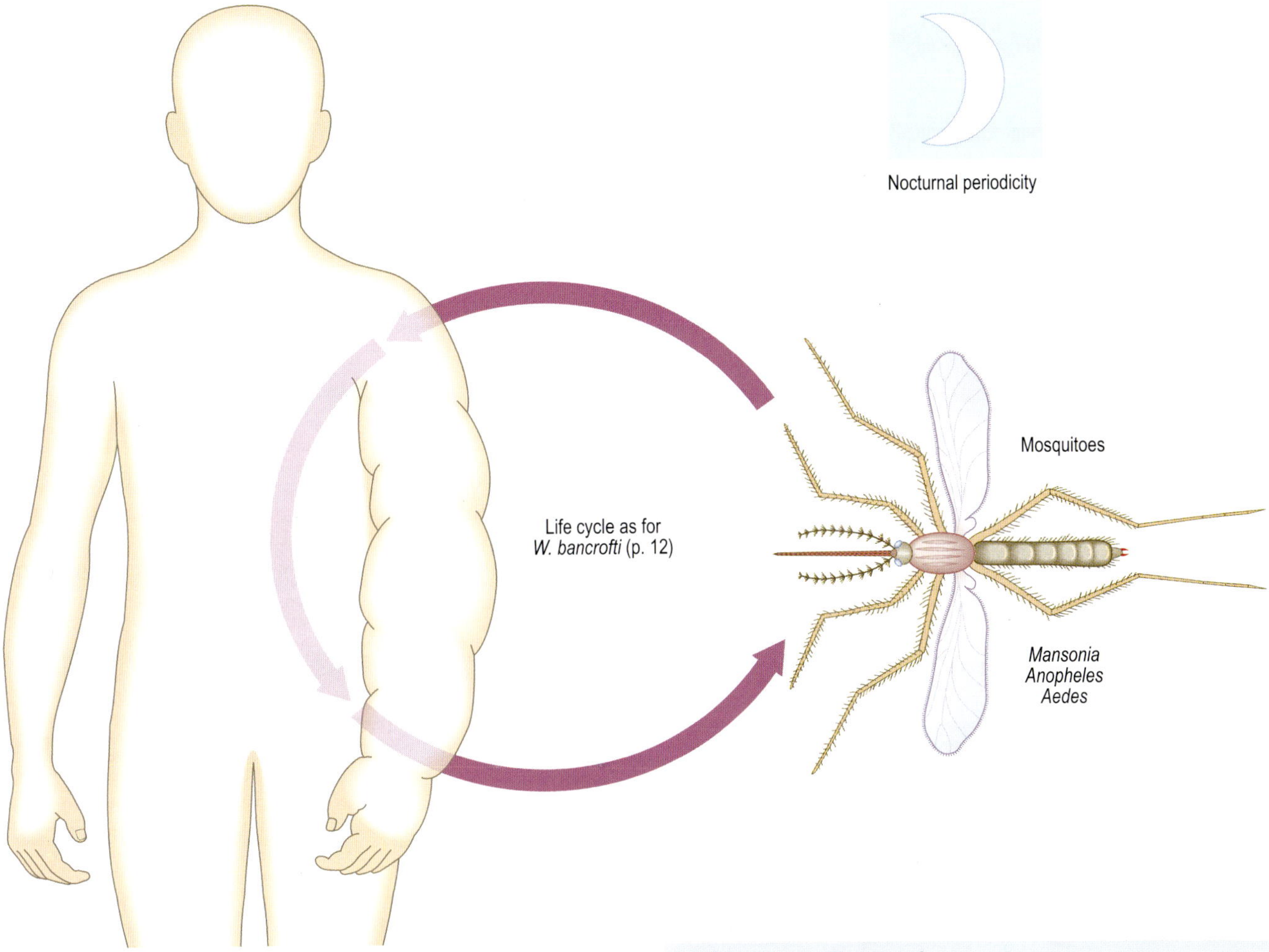

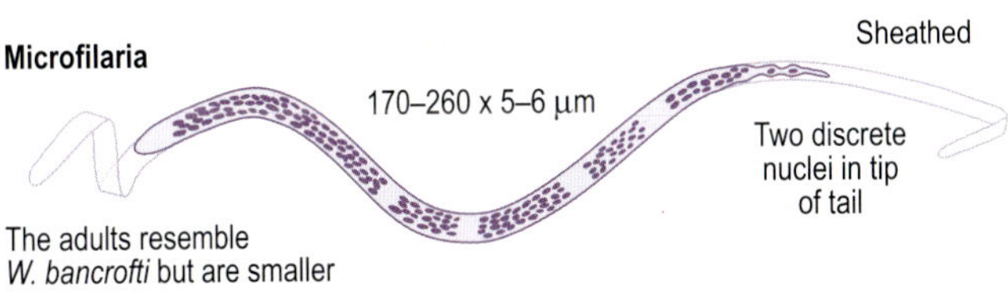

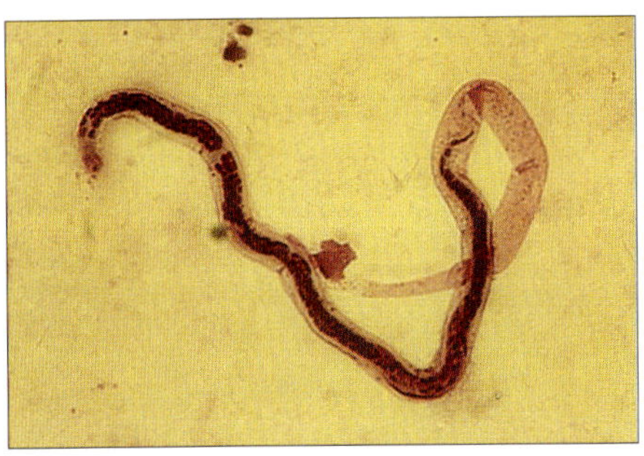

Pathology and Clinical features

These are similar to those of *Wuchereria*, but *Brugia* more commonly affects the upper limbs. Hydrocoele, other genital lesions and chyluria are rare.

Laboratory diagnosis

As for *Wuchereria bancrofti* except for the specific antigen test.

Distribution

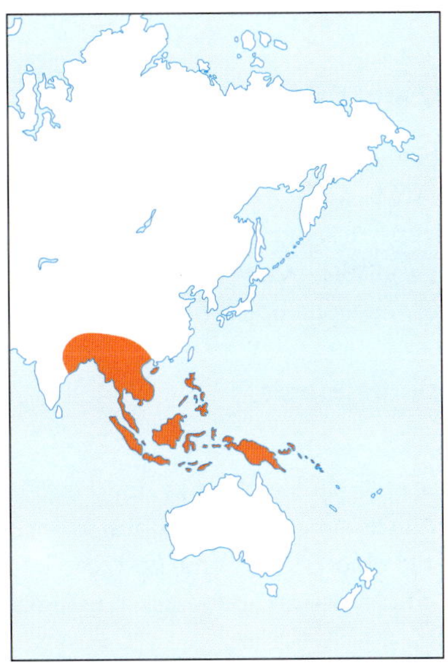

Loa loa (eye worm)

Life cycle

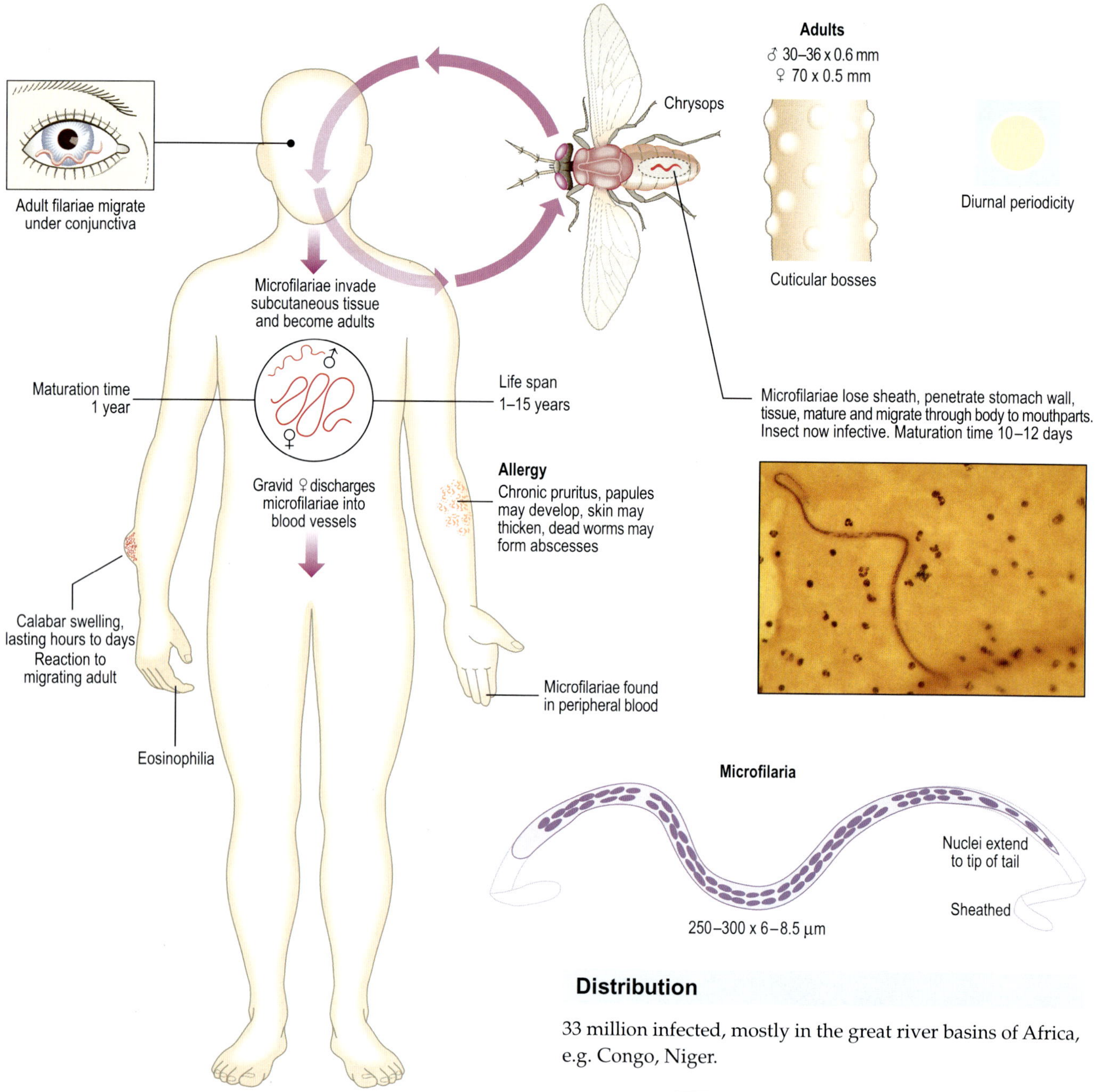

Distribution

33 million infected, mostly in the great river basins of Africa, e.g. Congo, Niger.

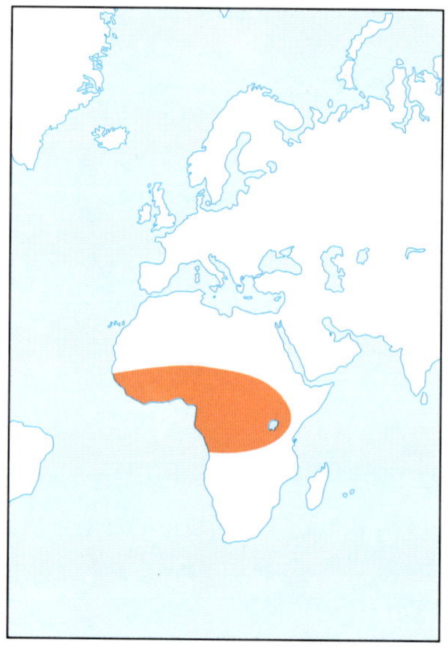

Pathology and Clinical features

Transient subcutaneous (Calabar) swellings due to hypersensitivity to adult excretory products.

The adult worm may appear under the conjunctiva and can be removed surgically. Symptoms include fatigue, chronic pruritus, rarely encephalopathy or nephropathy.

Laboratory diagnosis

Eosinophilia.

Microfilariae are found in blood by day (between noon and 14:00 hours). Nuclepore membrane filtration or centrifugation after lysis of the blood (Knott's method) can be used.

Serology. ELISA detects antibodies to filarial antigens but is non-specific.

Onchocerca volvulus (blinding worm)

Life cycle

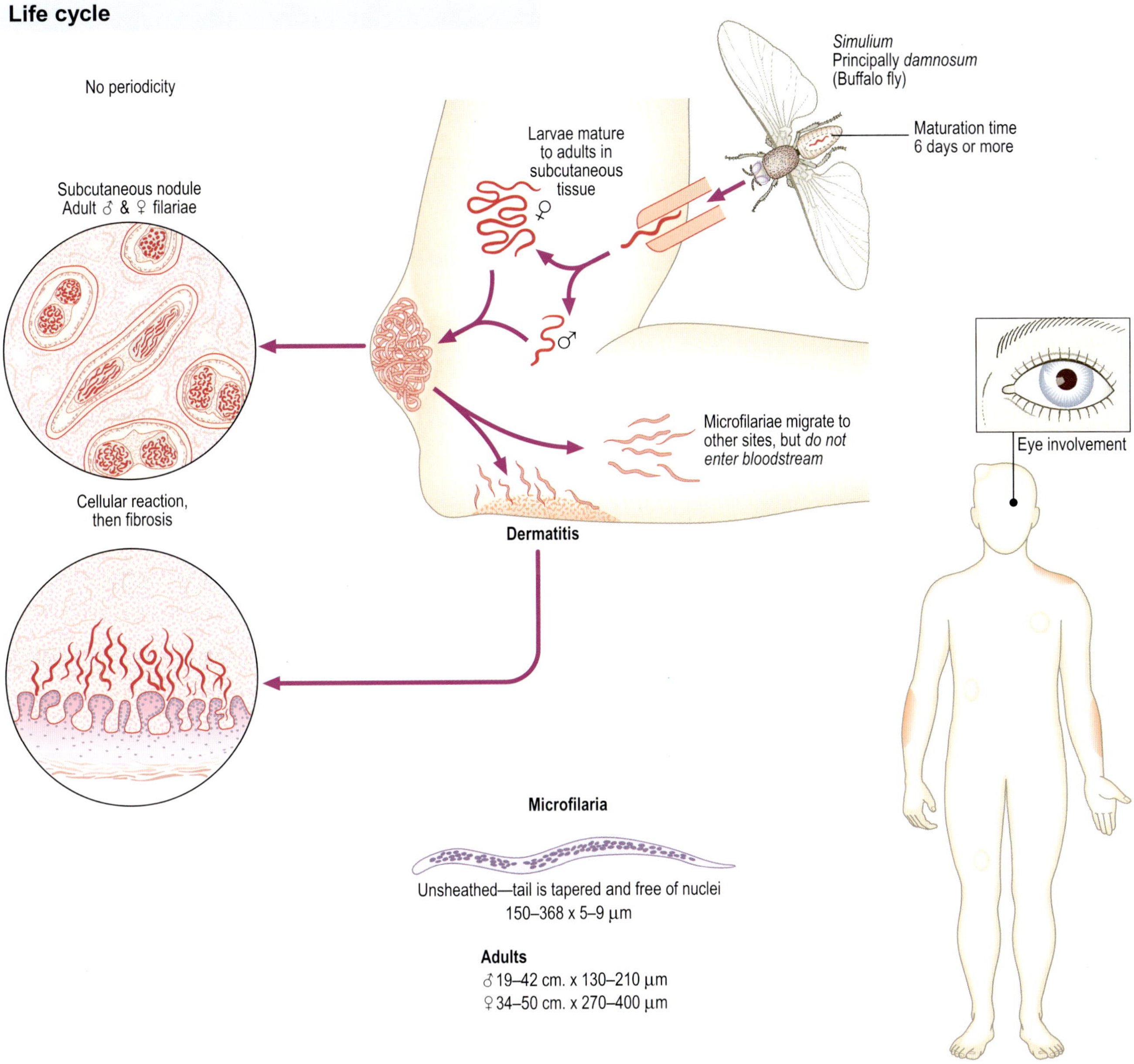

Pathology and Clinical features

Fibrous nodules develop round the adult worms, especially over the iliac crests. There may be some lymphatic obstruction; elephantiasis has been noted in Africa. The microfilariae cause itching, excoriation, urticaria, depigmentation, lichenification, 'sowda' and lymphadenopathy. When invading the eye, they can cause inflammatory lesions in any part of the eye such as sclerosing keratitis, choroidoretinitis and optic atrophy. Blindness may ensue.

Where microfilariae cannot be demonstrated, a Mazzotti test (DEC provocation test) can be useful.

Laboratory diagnosis

Eosinophilia.

Adult worms can be detected in excised nodules, microfilariae in the anterior chamber of the eye (slit lamp), skin snips and rarely in blood and urine.

Specific serodiagnosis by ELISA and PCR for parasite DNA on skin samples is in use.

Distribution

17 million infected worldwide.

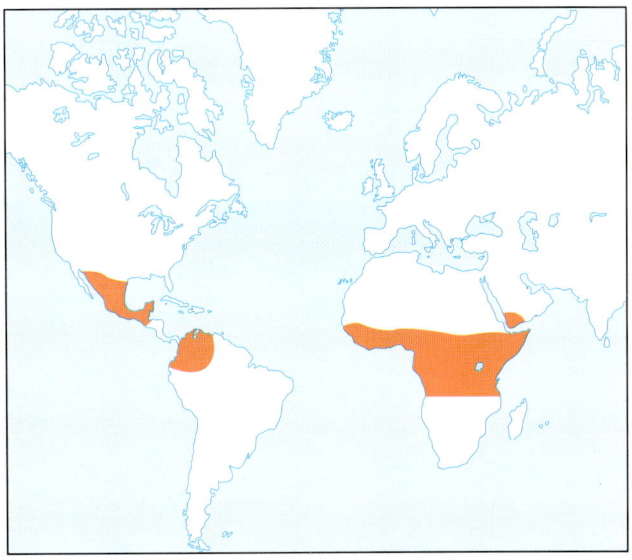

Other filarial worms

These worms are much less pathogenic. Microfilariae of other species are unsheathed, may be found in the blood and tissues and differentiation from *Wuchereria* and *Brugia* is necessary.

Filtration requires 3 micron pore size membrane, because of the smaller size of these microfilariae.

No periodicity.

Mansonella perstans

Found in Tropical Africa and the coasts of Central and South America. The vector is the midge *Culicoides*. Microfilariae can be found in the blood.

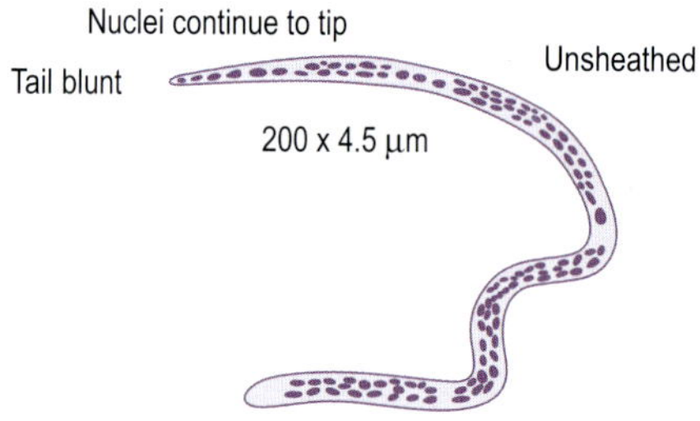

Mansonella streptocerca

Found in Africa. The vector is the midge *Culicoides*. Microfilariae can be found in the skin.

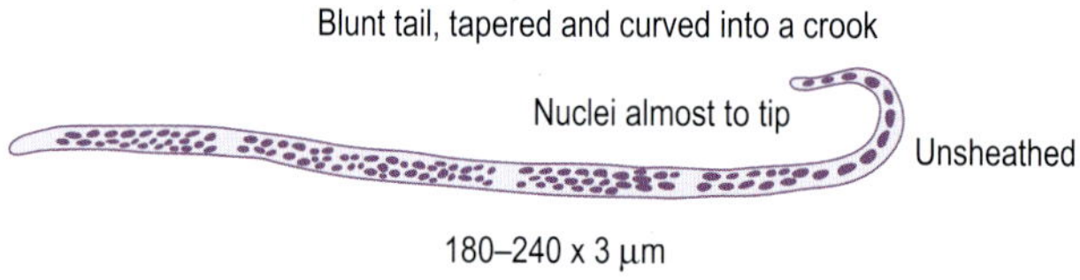

Mansonella ozzardi

Found in South America and the Caribbean. The vector is the midge *Culicoides*.

Microfilariae can be found in the blood and skin.

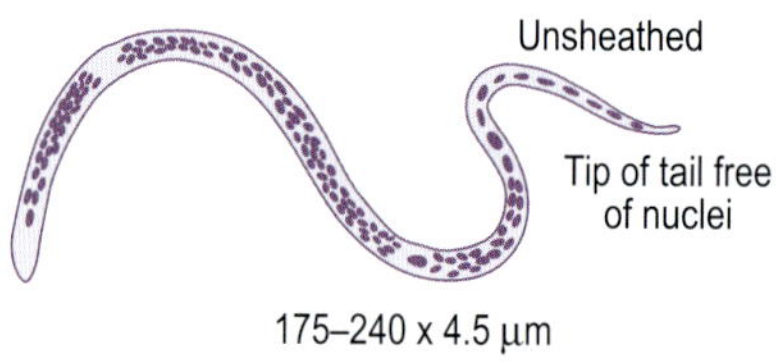

Dracunculus medinensis (Guinea worm)

Life cycle

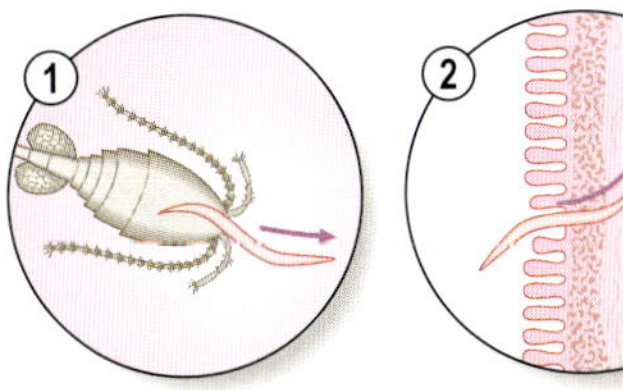

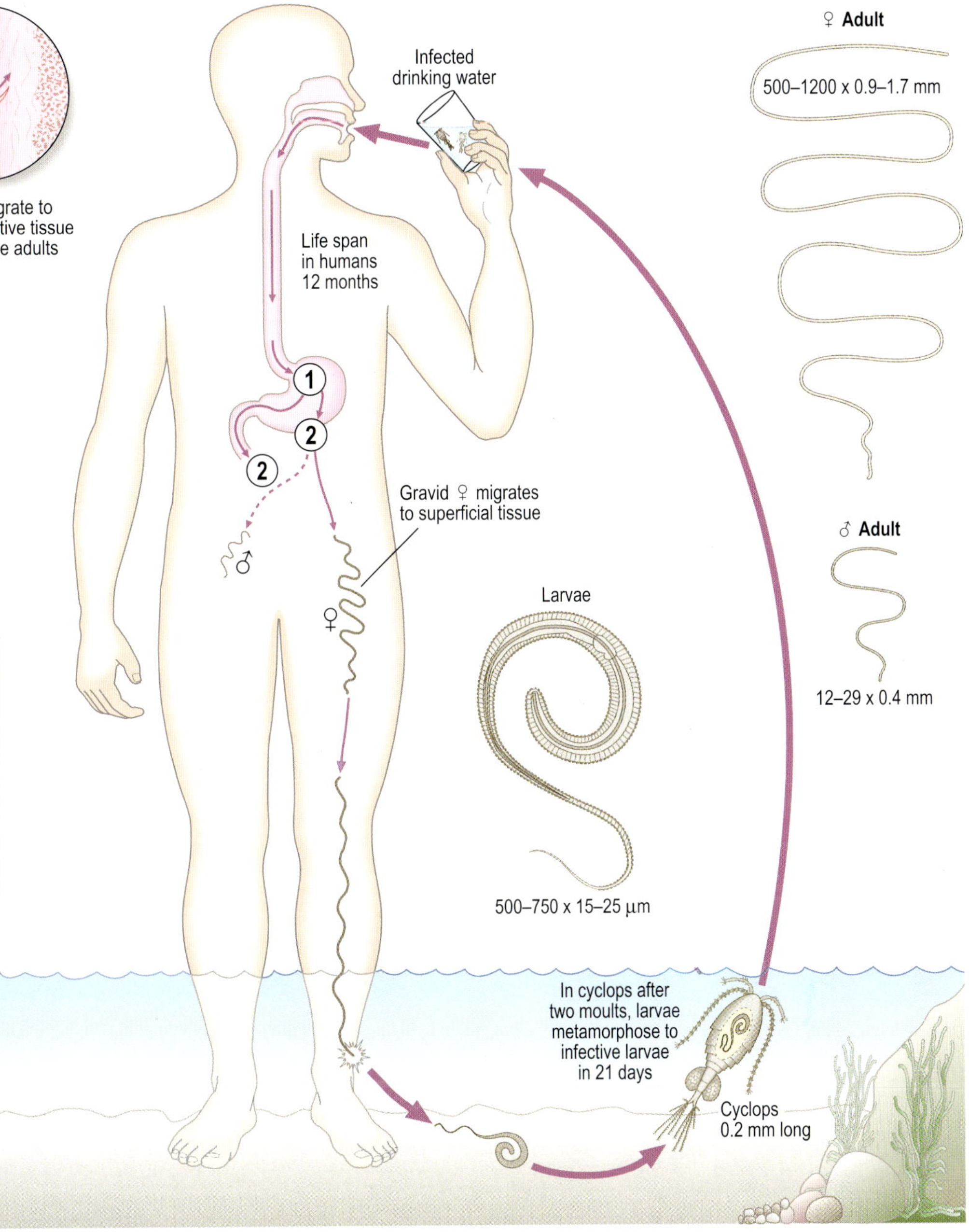

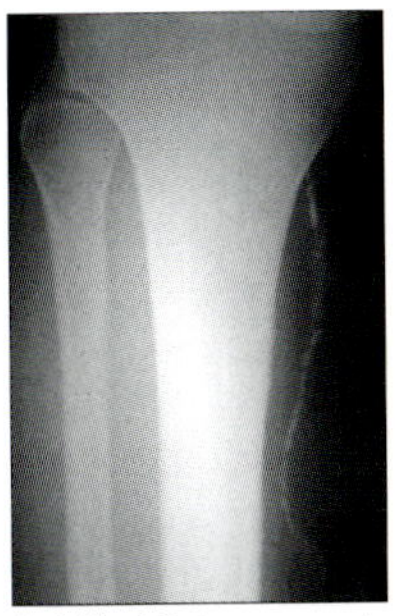

Calcified Guinea worm in X-ray of leg

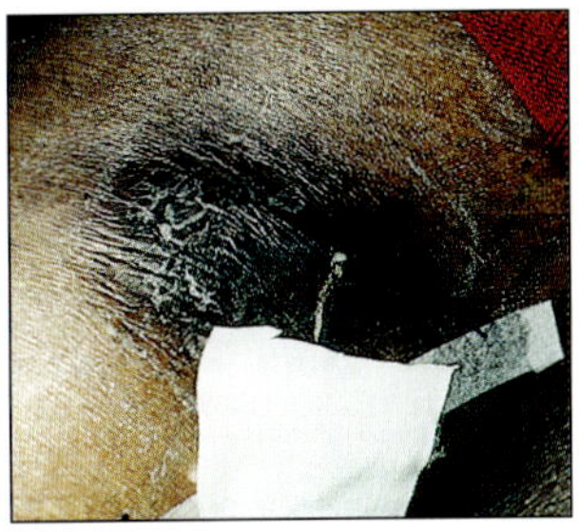

Guinea worm

Pathology and Clinical features

The gravid female causes itching, urticaria and a burning sensation. A blister appears which bursts to become an ulcer (usually leg) with discharge of embryos and some fibrosis. The adult female may be seen protruding from the ulcer. There is often secondary bacterial infection, and sometimes arthritis of the knee and ankle. Worms may fail to emerge, die and calcify.

Laboratory diagnosis

Eosinophilia.

Larvae may be found in fluid from the ulcer.

Distribution

70 000 infected worldwide.

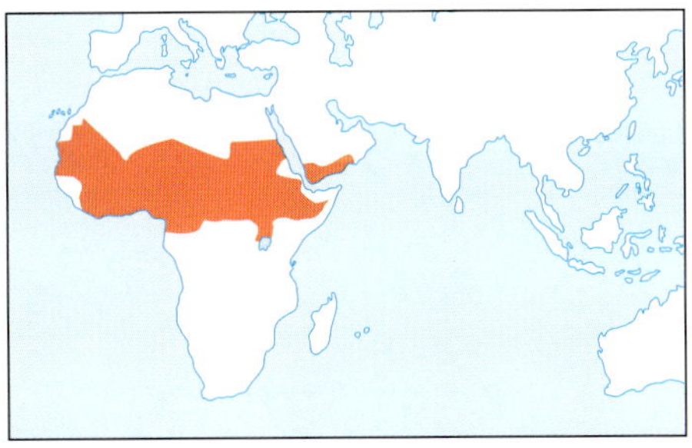

Areas where dracunculiasis is endemic (based on reported cases in 1997). (Map reprinted from Weekly Epidemiological Record 1997; 72(6):33-35; prepared by WHO/UNICEF HealthMap Programme & CTD/DRA, Geneva: WHO.)

Phasmid Nematodes

Toxocara canis (dog round worm)

Morphology

Toxocara: body is bent ventrally. *Toxascaris*: body is bent dorsally.

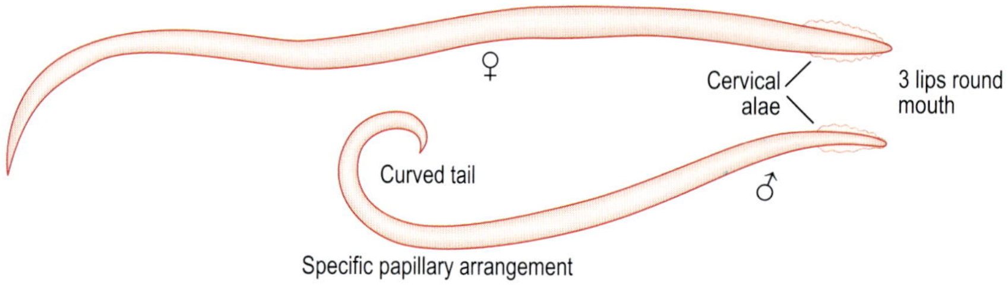

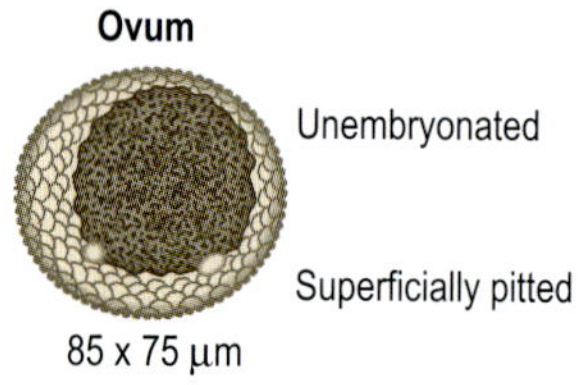

Life cycle and occurrence

Ocular larva migrans (OLM) and visceral larva migrans (VLM) usually occur as distinct entities without overlap. VLM occurs in younger children and gives rise to fever, pneumonitis and hepatomegaly. Myocarditis, convulsions, psychiatric changes or encephalopathy may occur. OLM presents as unilateral visual loss, often with squint. Retinal detachment, endophthalmitis or papillitis may occur.

Laboratory diagnosis

Eosinophilia.

Serology. Antibody detection by ELISA on serum. A vitreous sample may be required in OLM. Examination of environmental soil samples for ova by concentration techniques may be an aid to control.

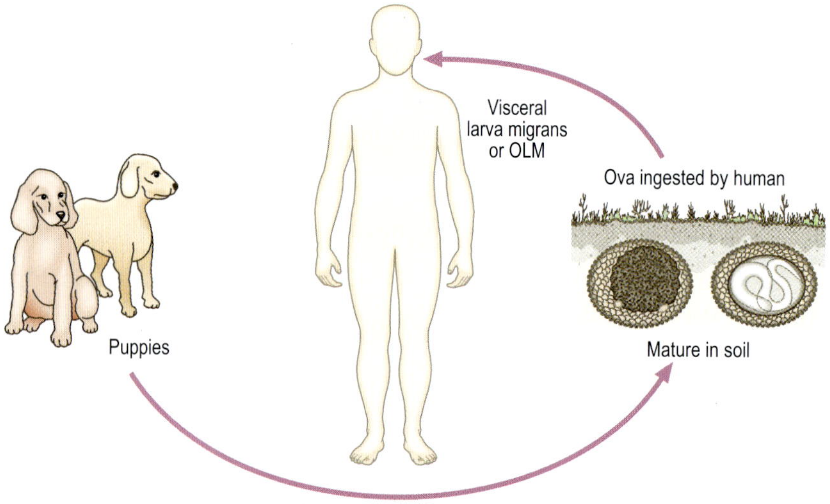

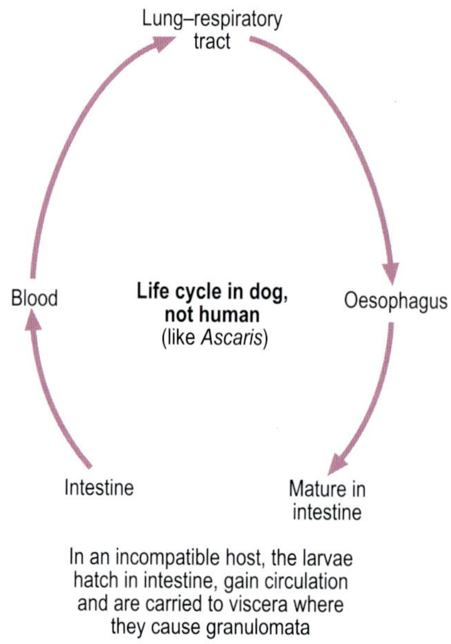

Toxocara cati (cat round worm)

Life cycle

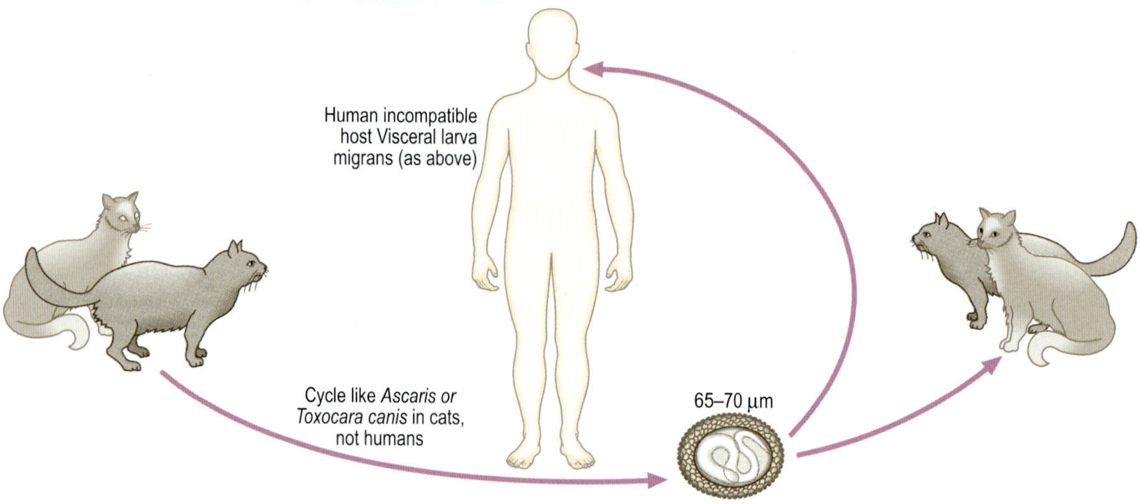

Gnathostoma spinigerum

Morphology

Stout, reddish-coloured worms

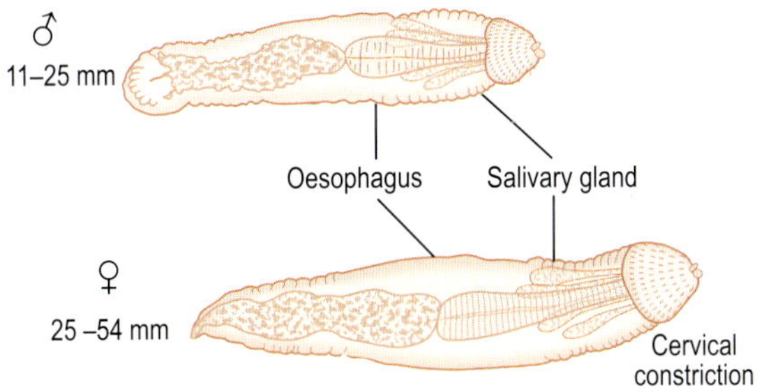

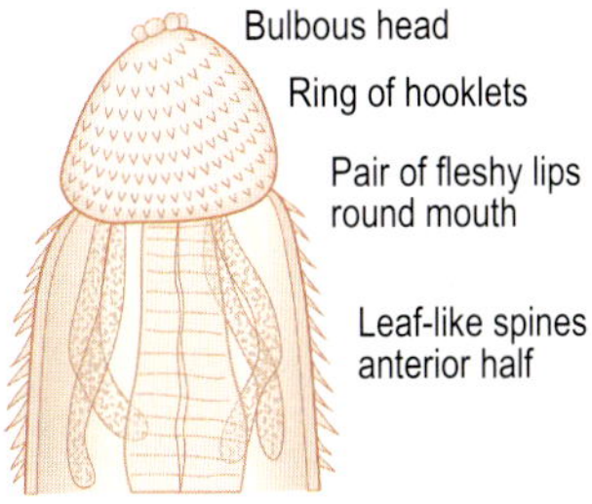

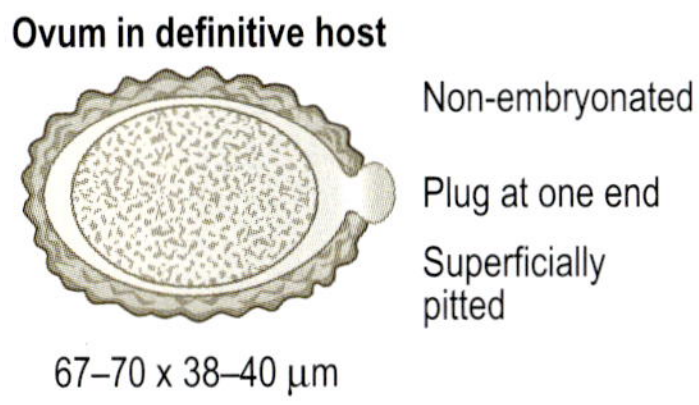

Life cycle and occurrence

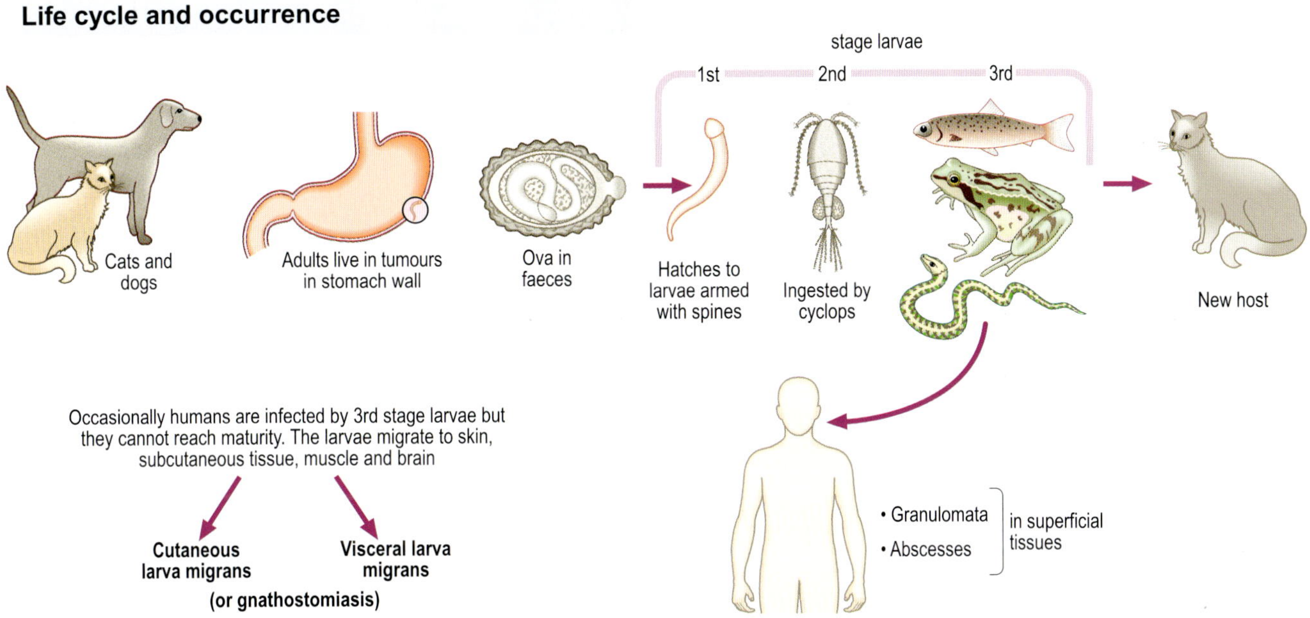

Laboratory diagnosis

ELISA for antibody detection. Histology or mophology of worm if excised.

Distribution

South East Asia, mainly Thailand.

Cutaneous Larva Migrans (creeping eruption)

Caused by non-human hookworm larvae.

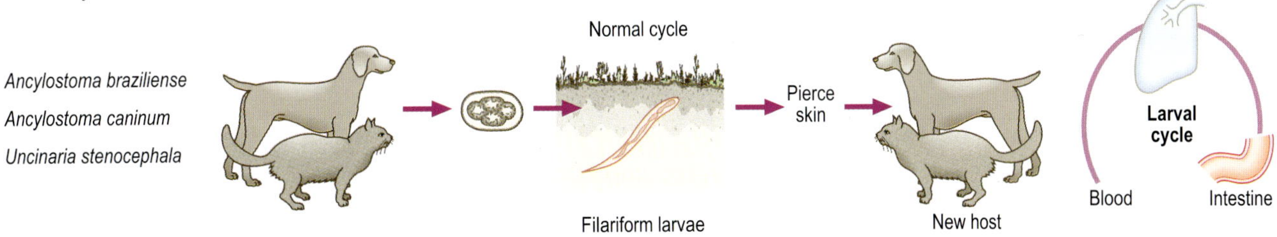

If they successfully invade humans, the intensely itchy infection lasts for months.

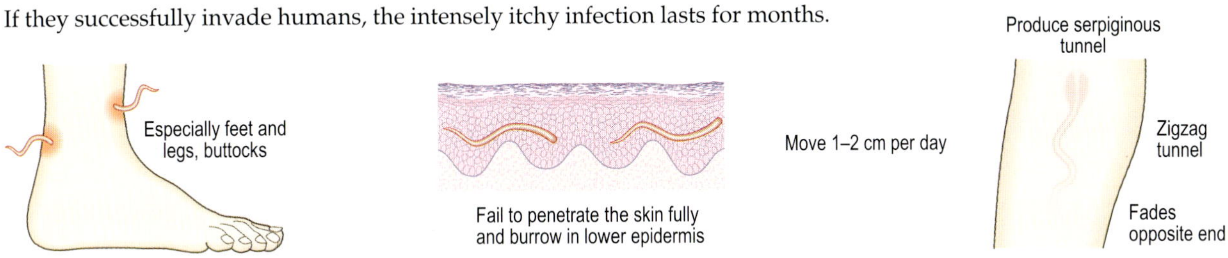

Cestode (tape) worms

Taenia solium (pork tape worm)

Life cycle

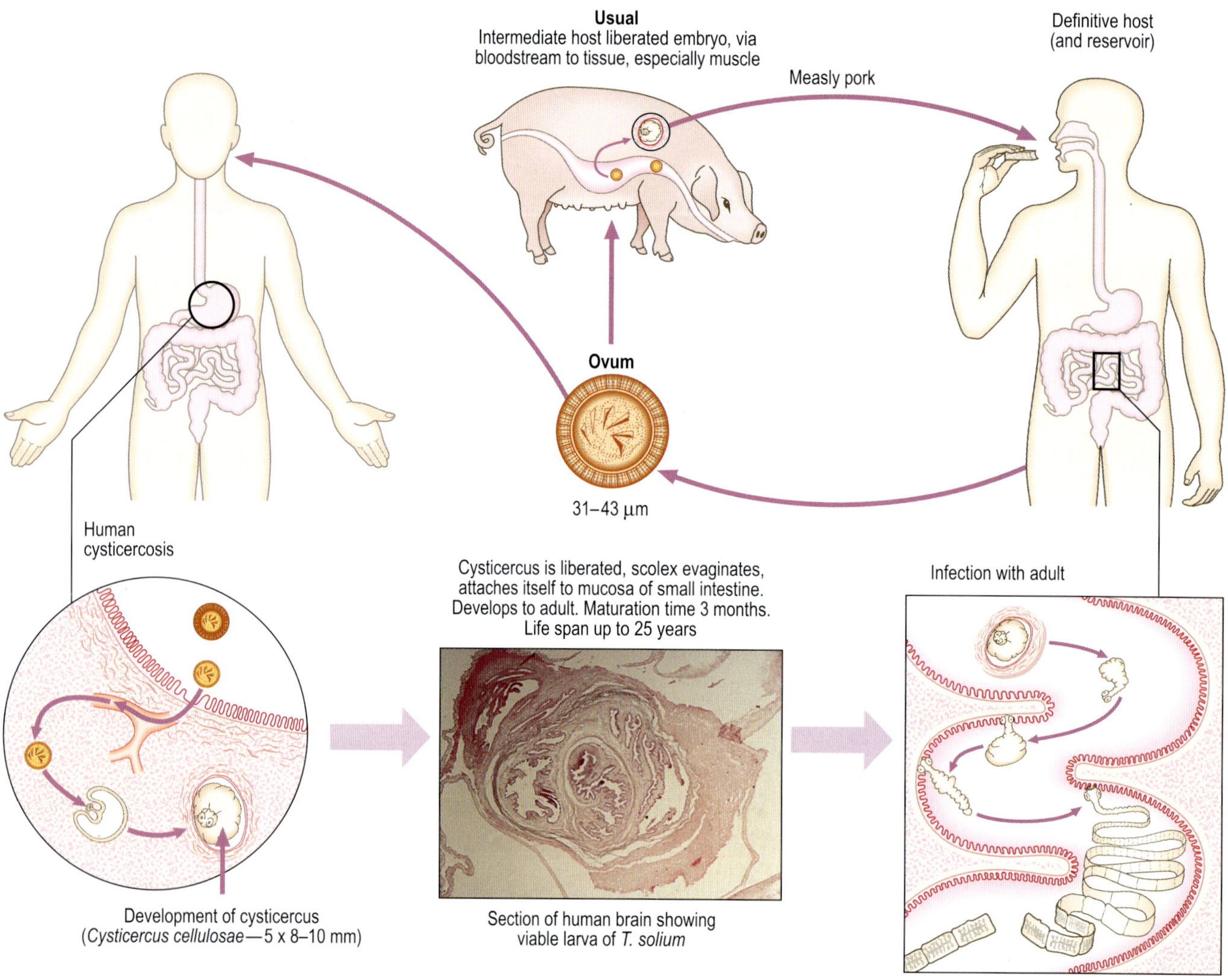

Pathology and Clinical features

Infection by larvae (cysticercosis). Cysticerci, generally multiple, may occur in any site but are more frequent in the brain and muscle. They excite reaction in the area, especially when they die, which manifests as inflammation, fibrosis and later some calcification. This leads to focal CNS syndromes, especially epilepsy.

Infection with adults. Often there can be no pathology, but there might be mild irritation of intestinal mucosa.

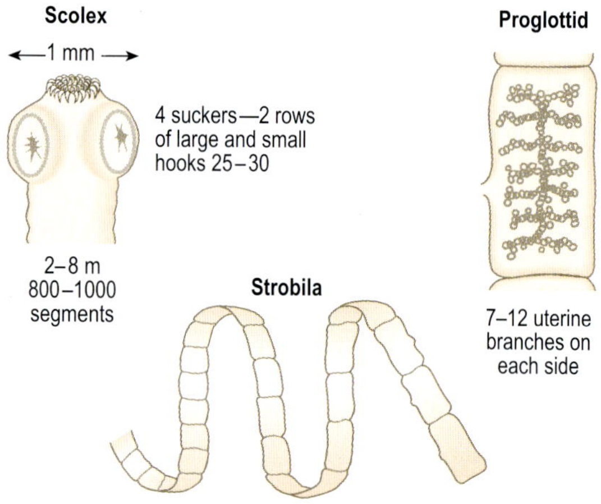

Laboratory diagnosis

Eosinophilia.

Larval infections. There are several methods, including histological examination of biopsy material, serology (IFAT, ELISA, EITB) and radiology (CT or MRI scan of the brain, X-ray of the thigh muscles).

Pure infection with the adult. Gravid segments, ova and scolex can be found in faeces. The uterine branches of the mature segments can be demonstrated by injection of Indian ink through the uterine pore.

Distribution

5 million people infected worldwide. *Taenia solium* is endemic in pig-rearing areas of the world where hygiene and animal husbandry are poor.

Taenia saginata (beef tape worm)

Life cycle

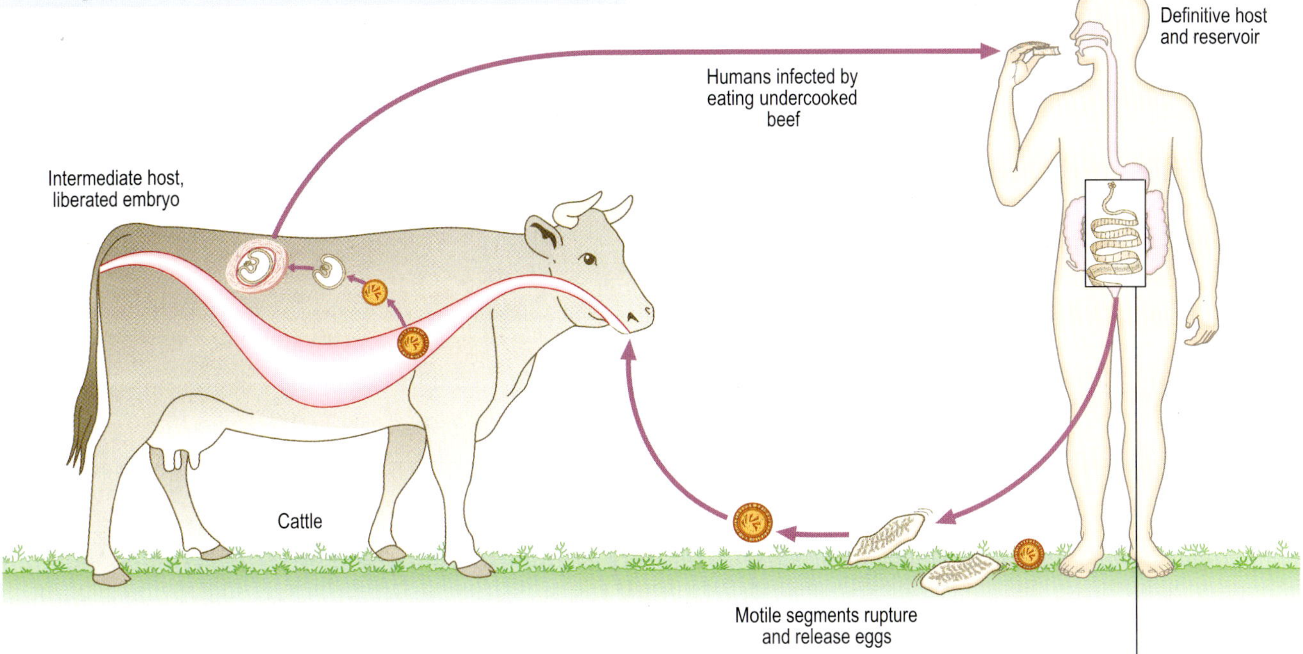

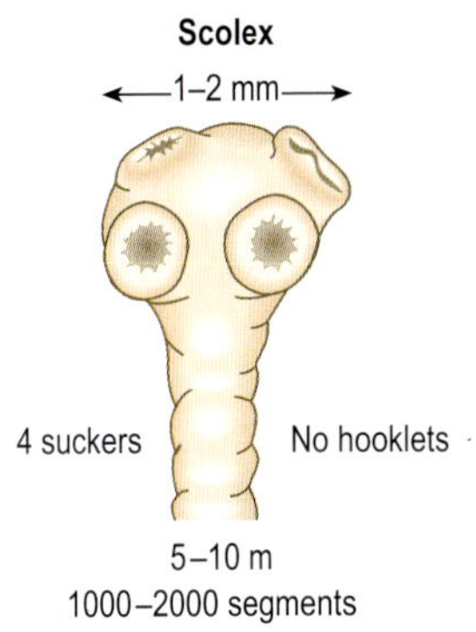

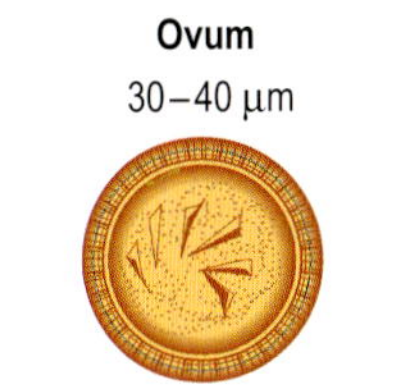

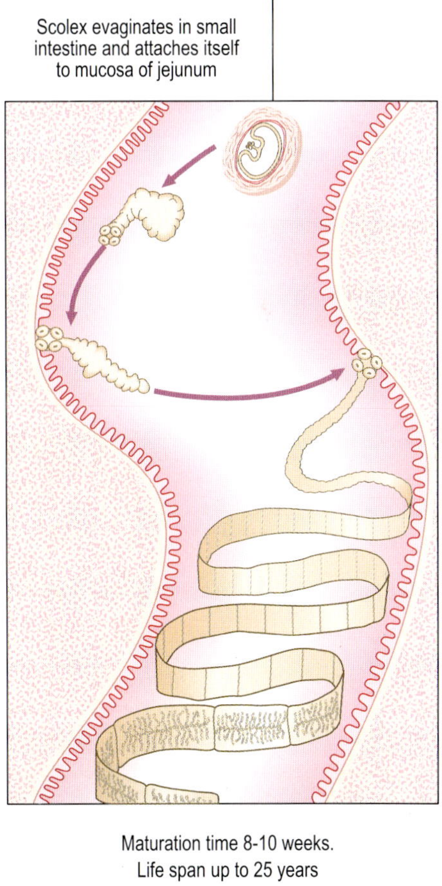

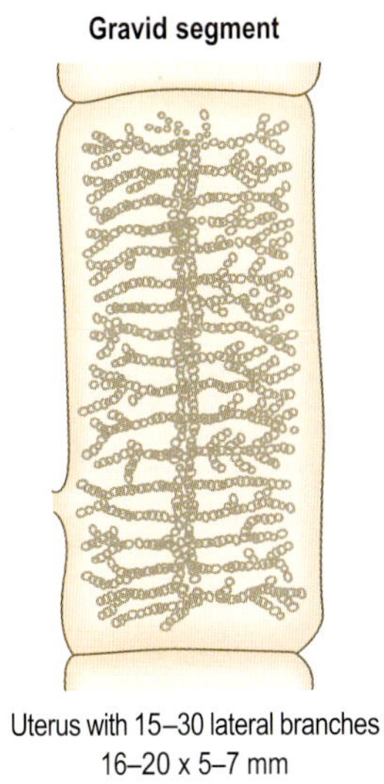

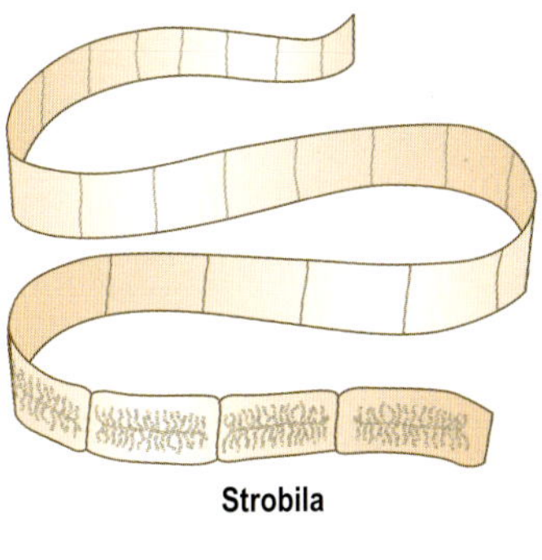

Pathology and Clinical features

Usually there is no pathology as *Cysticercus bovis* is unknown in humans. Occasionally there is vague alimentary upset.

Laboratory diagnosis

Gravid segments, ova and scolex can be found in faeces. Uterine branches of the mature segments may be seen in a crush preparation between two glass slides, or by Indian ink preparation, as in *T. solium*. Ova are also found on the perianal skin (on clear adhesive tape slides).

Distribution

Taenia saginata is found in beef-eating areas, especially in the tropics.

Dwarf tape worms

Hymenolepis nana

Life cycle

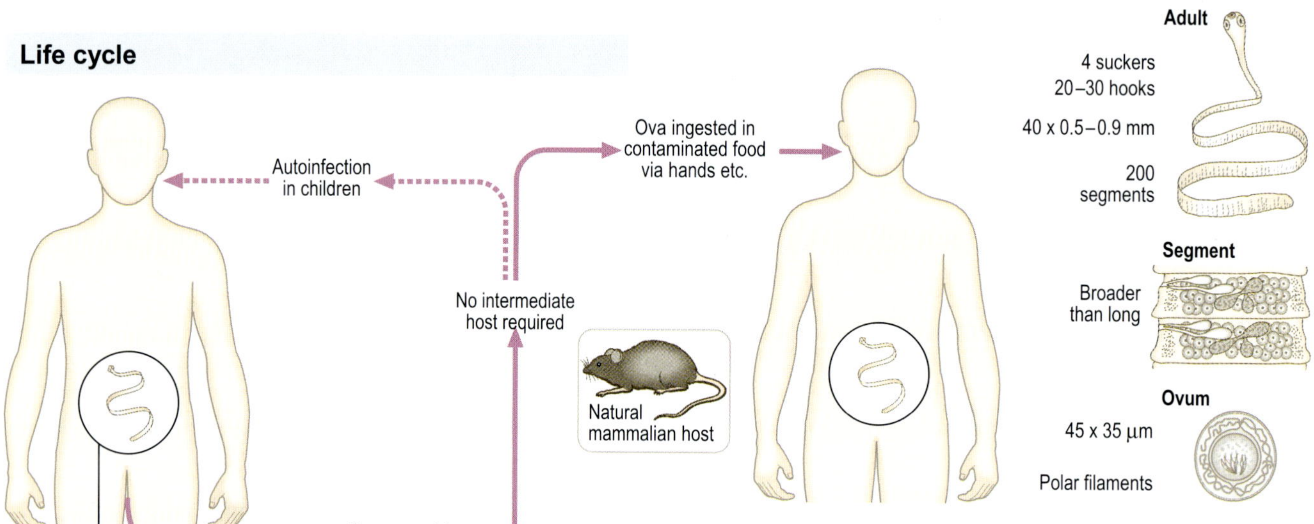

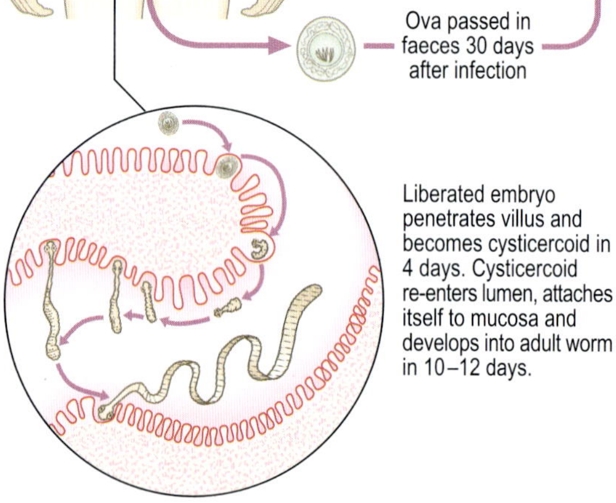

Pathology and Clinical features

Often there are none, but with heavy infection there may be abdominal pain and diarrhoea. Anaemia and nervous symptoms, including dizziness and irritability, can occur in children.

Laboratory diagnosis

Eosinophilia may be present. Ova found in faeces.

Distribution

36 million people are infected worldwide.

Hymenolepis diminuta (rat tape worm)

Life cycle

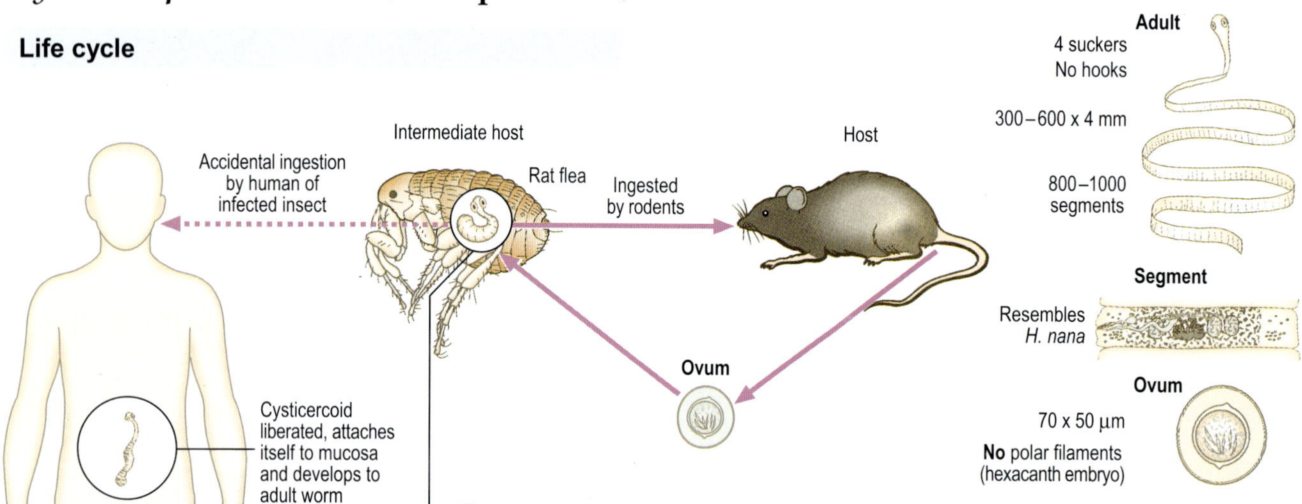

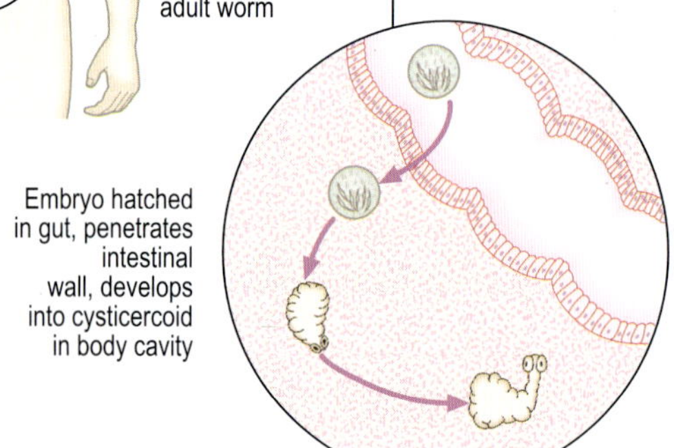

Pathology and Clinical features

Generally there is no effect on the host.

Laboratory diagnosis

Ova in faeces.

Distribution

Worldwide, but rare in humans.

Diphyllobothrium latum (fish tape worm)

Life cycle

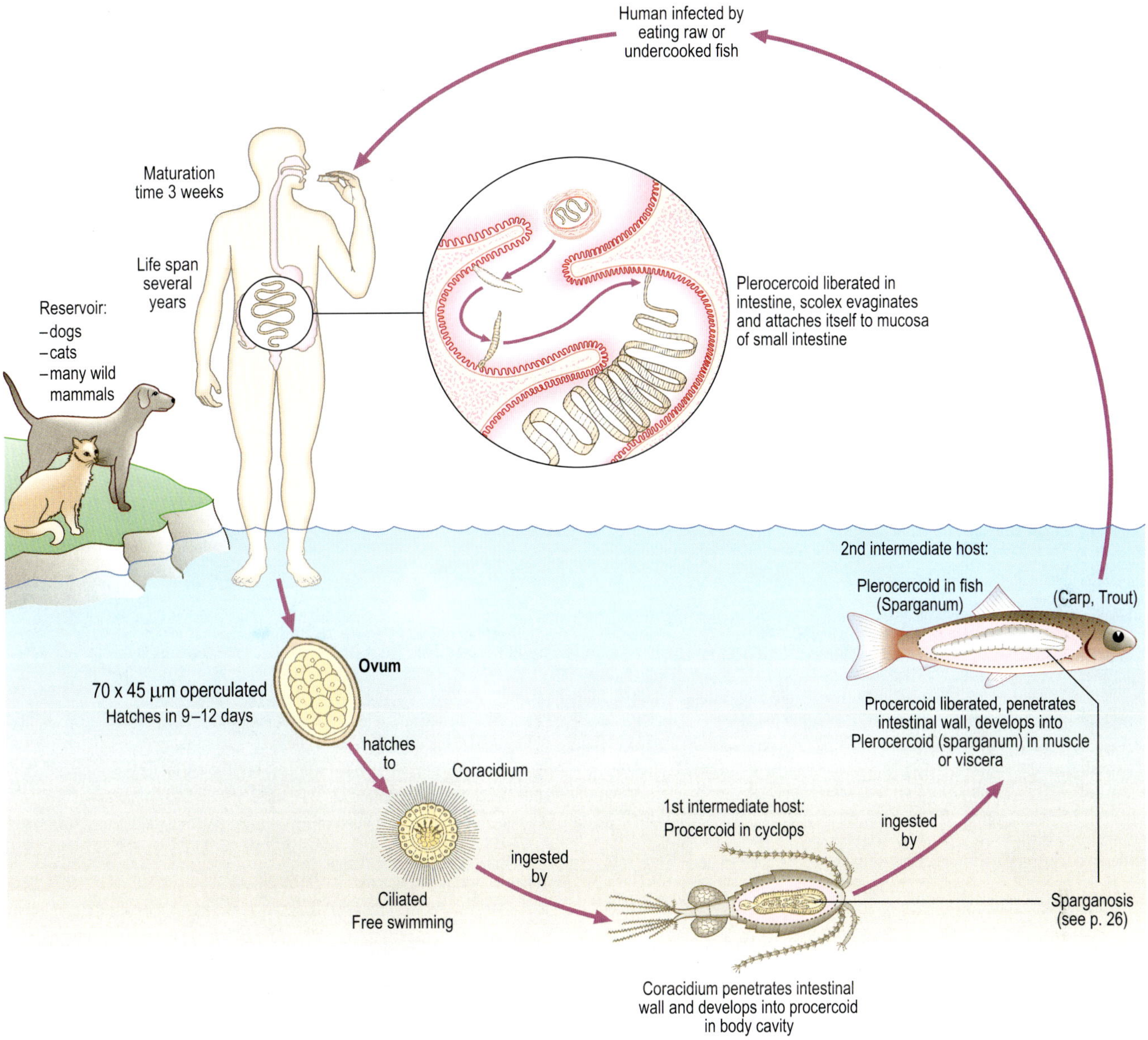

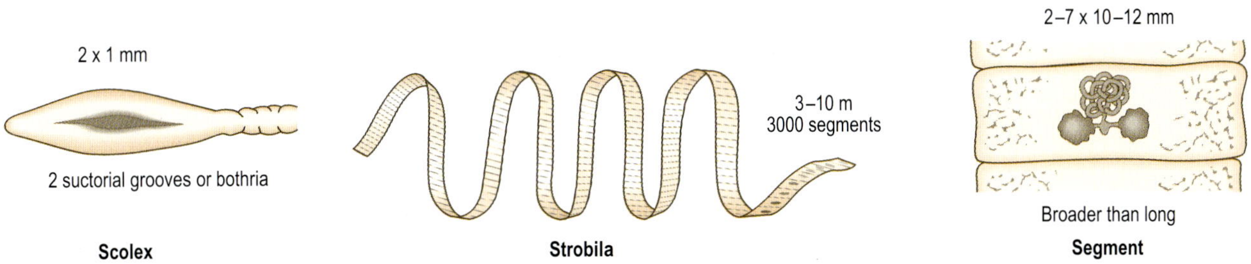

Pathology and Clinical features

Generally there is none, but occasionally there can be megaloblastic anaemia (through absorption of vitamin B_{12} by the worm).

Laboratory diagnosis

Eggs and gravid segments can appear in faeces. Megaloblastic anaemia (low serum B_{12}).

Distribution

16 million infected worldwide in eastern seaboard of Canada and America, Brazil, Baltic States, parts of West Africa, North Siberia and South East Asia.

Larval forms of cestode infection in humans

Sparganosis

Life cycle of such tapeworms

Sparganosis is caused by the extra-intestinal presence in the human body of larvae of non-human tapeworms of the genus *Spirometra*.

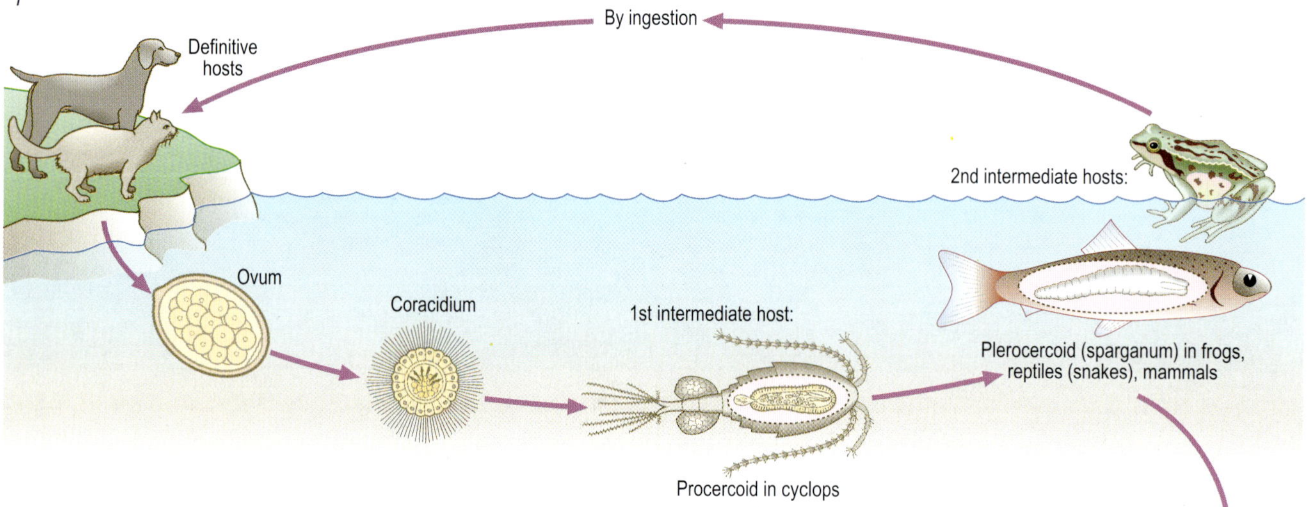

Human sparganosis by

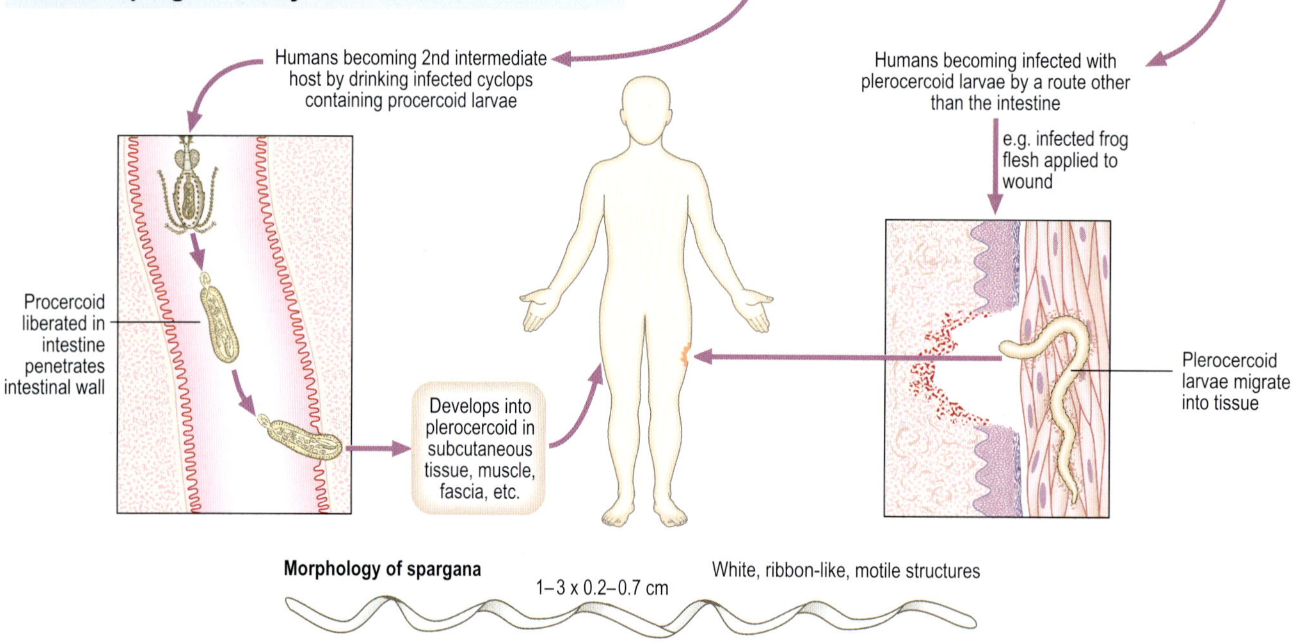

Pathology and Clinical features

Infestation with living larvae causes a painful oedematous reaction. Dead larvae cause intense local inflammatory reactions. There are numerous eosinophils and there can be abscess formation. There can be ocular sparganosis in the soft tissues near the eye, resulting in severe damage. Invasion of the CNS may occur.

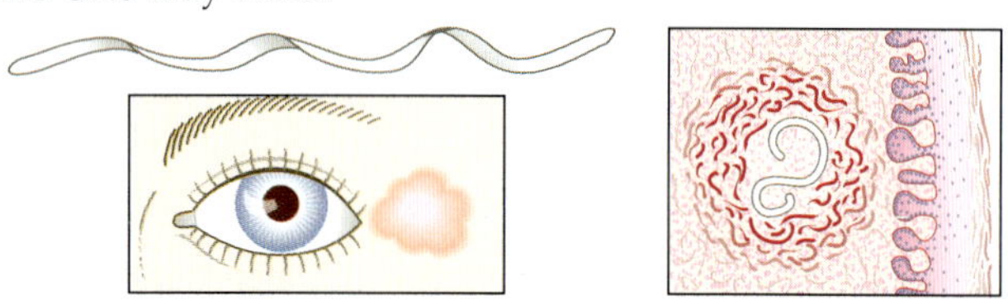

Types of spargana

Most Spargana do not proliferate in human tissues. *Sparganum proliferum* is a very rare parasite in which sparganum proliferates by lateral budding.

Laboratory diagnosis

Diagnosis of the disease is by examination of biopsy material or excised larvae.

Distribution

The Far East mainly but occasionally elsewhere.

Echinococcus granulosus (dog tape worm)

Life cycle

Echinococcus granulosus causes hydatid disease.

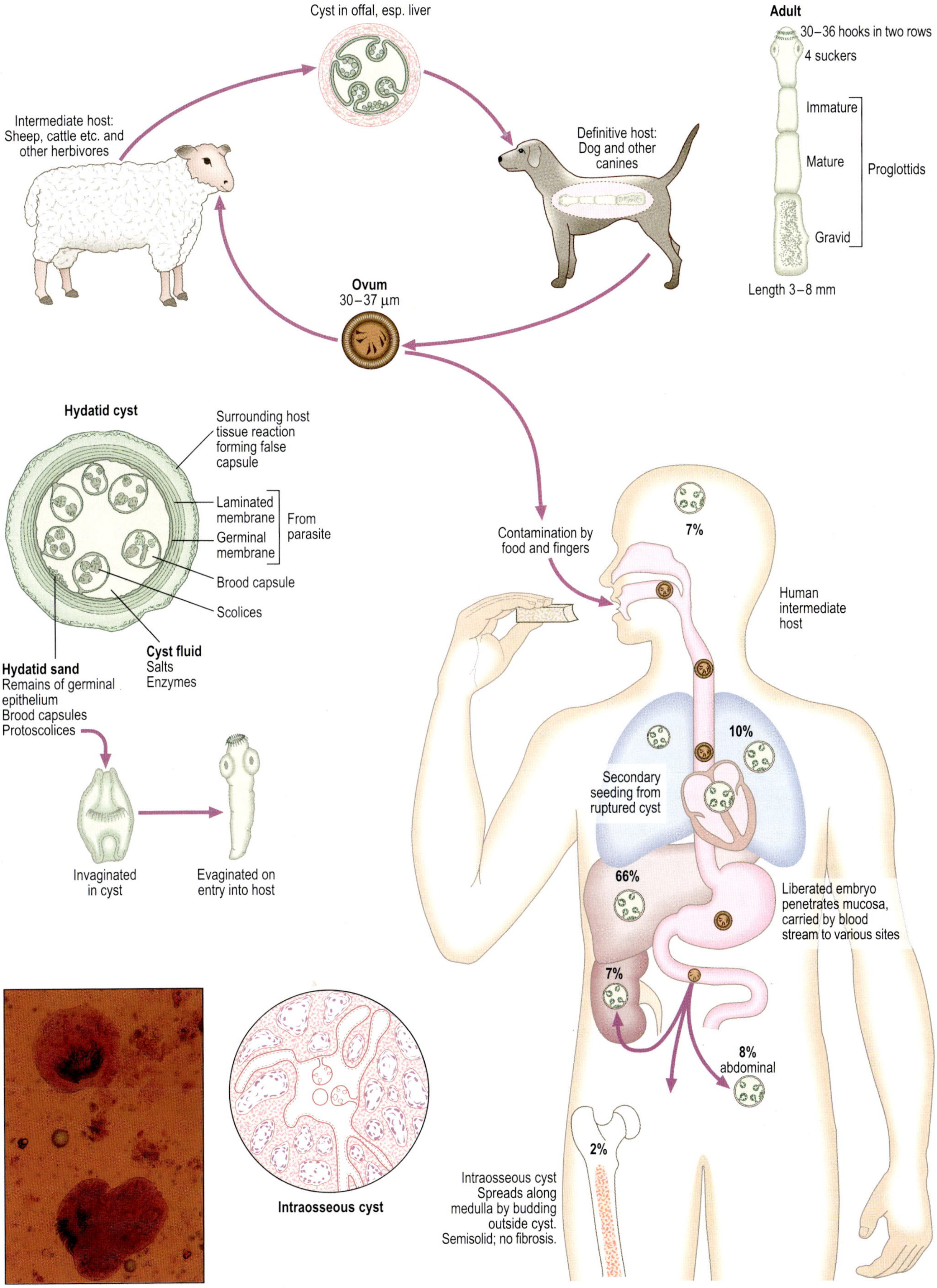

Echinococcus multilocularis

Life cycle

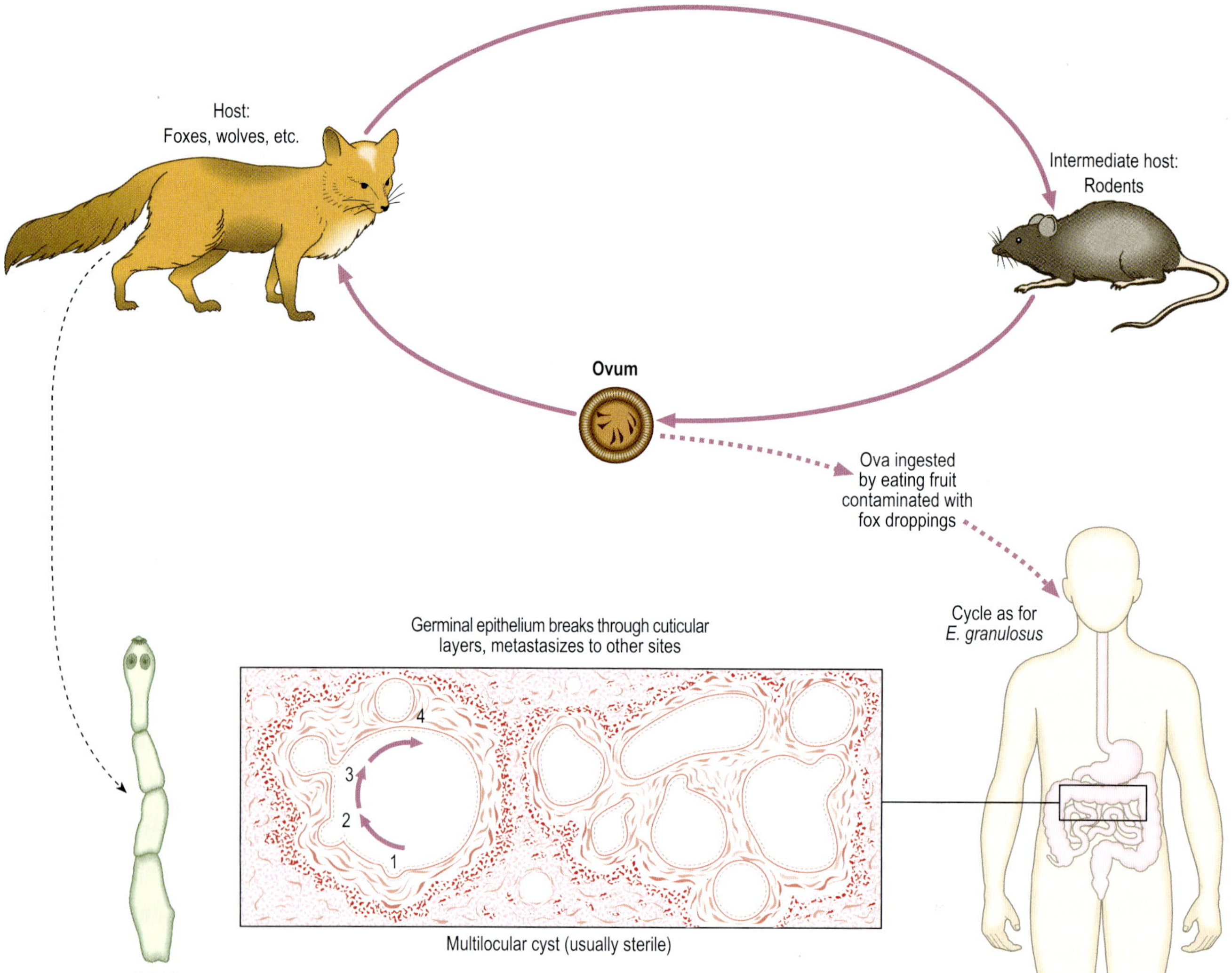

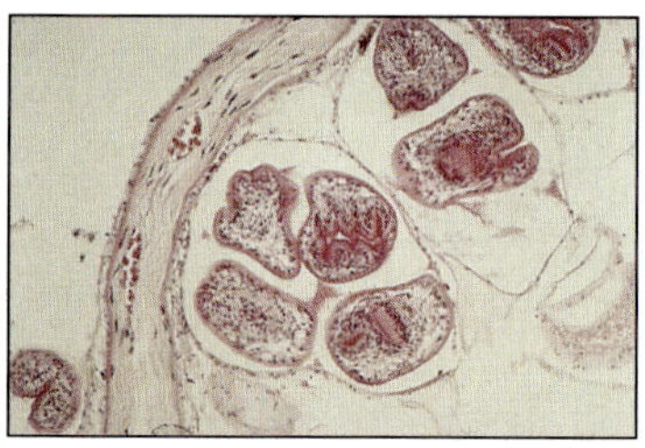

Pathology and Clinical features of hydatid disease

Echinococcus granulosus

Unilocular cysts. There is usually surrounding inflammatory reaction and fibrosis. After years, the cyst may die, shrink and calcify. There is general allergic reaction with eosinophilia, bronchospasm, etc. Pressure effects can cause local tissue damage and obstruction of natural channels. Rupture or leakage of the cyst can accentuate the allergic reaction. There can be anaphylactic shock and sometimes secondary implantation, for example in the peritoneal region. There can also be secondary infection with formation of abscess.

Osseus cysts. Usually there is no fibrosis although there is some cellular infiltration. Destruction of the bone can sometimes lead to spontaneous fracture.

Echinococcus multilocularis

Alveolar cysts. There are local pressure effects and allergy. Germinal epithelium can act like a neoplasm with local infiltration or distant metastases.

Laboratory diagnosis of hydatid disease

Use serological tests on serum (e.g. ELISA, complement fixation, counter current immunoelectrophoresis for Arc 5 or immunoblot). Microscopy of cyst fluid from operative specimens can be used to assess viability of protoscolices. Histological examination of a removed specimen is another possibility.

Distribution

1 million infected worldwide. *E. multilocularis* is rare in humans, but occurs in Northern Europe, Asia, North America and Arctic regions. *E. granulosus* is widespread in sheep-rearing areas of the world. Eradication is well advanced in Australia and New Zealand.

Trematode (flat) worms

Schistosoma species (blood flukes)

Life cycle for all species

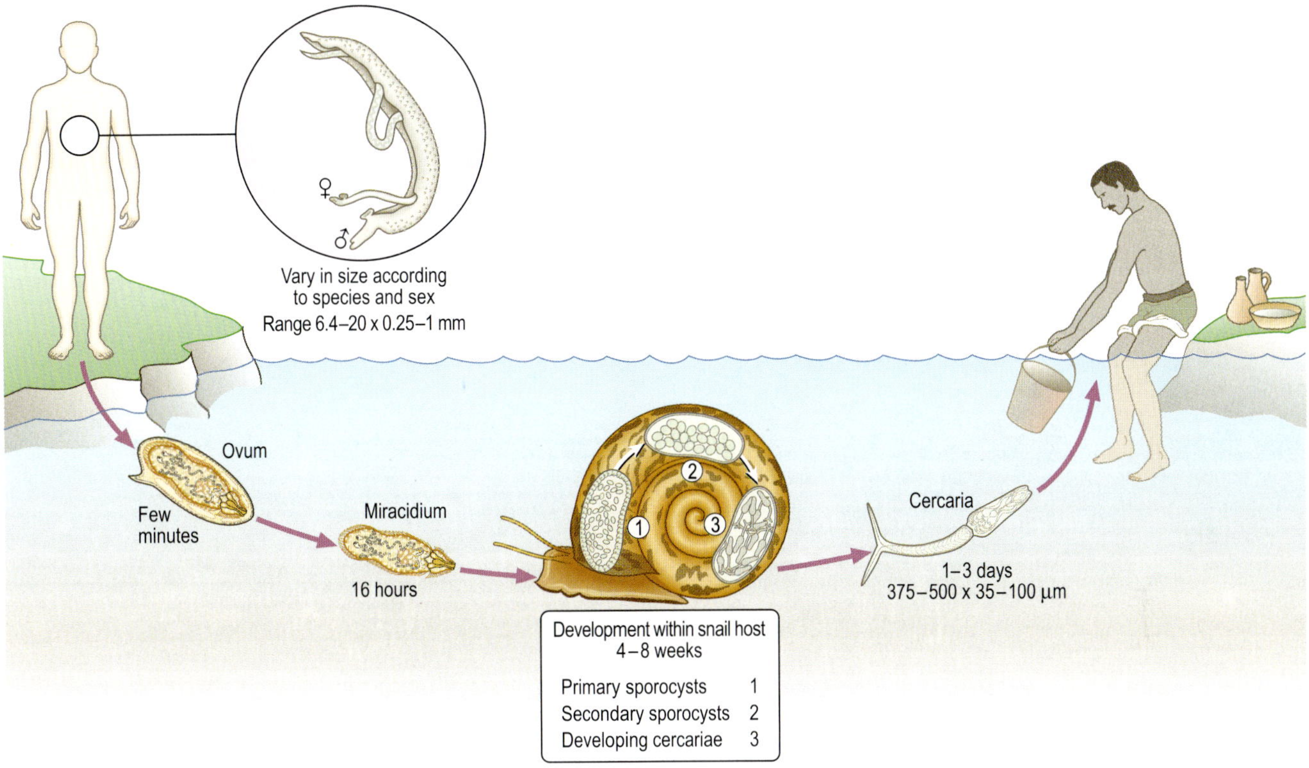

Life cycle in humans

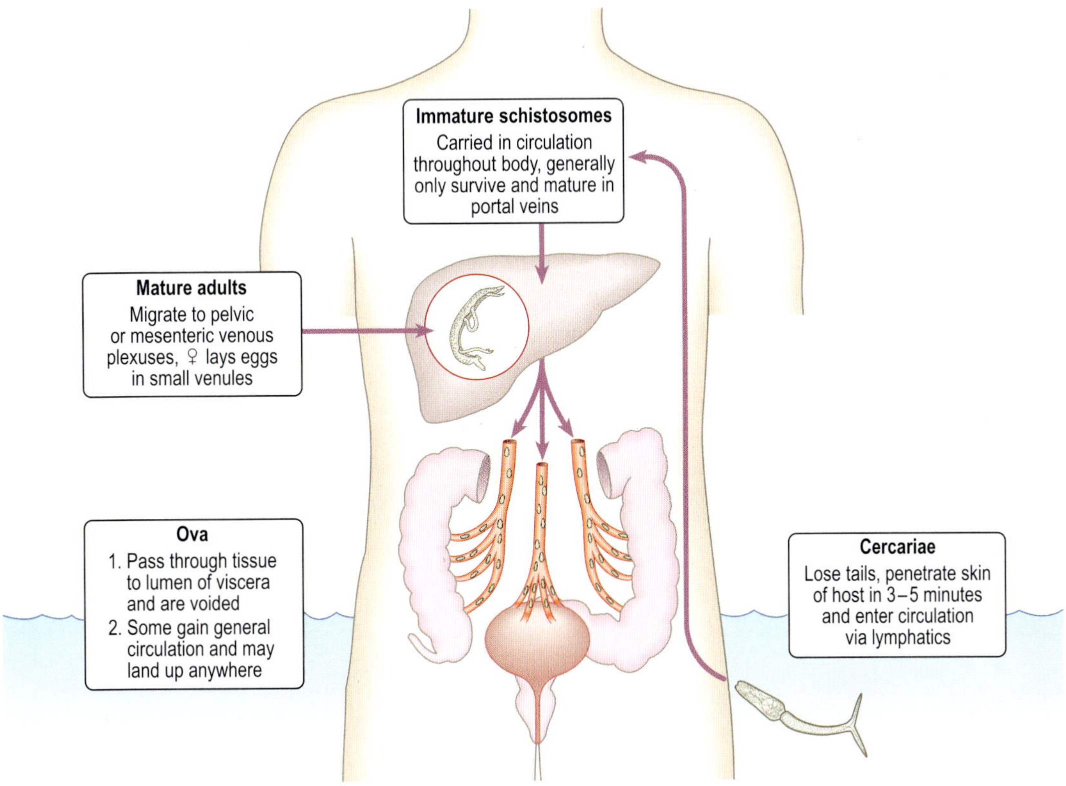

Schistosoma species (blood flukes) (Continued)

Morphology

S. haematobium

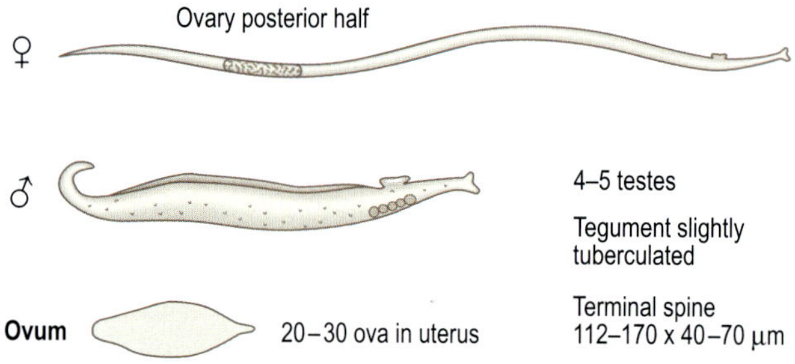

Host: *Bulinus*

S. mansoni

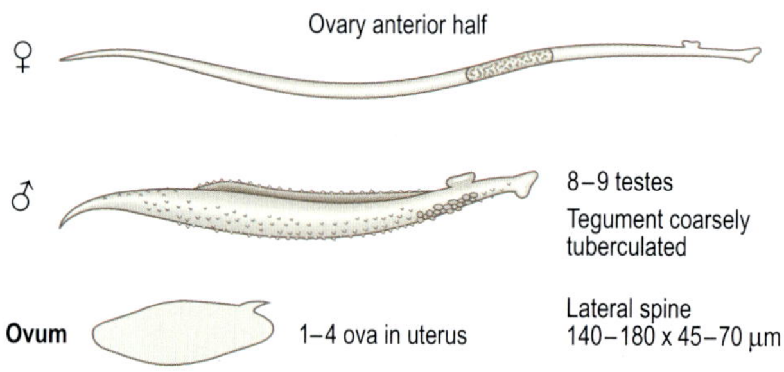

Host: *Biomphalaria*

S. japonicum

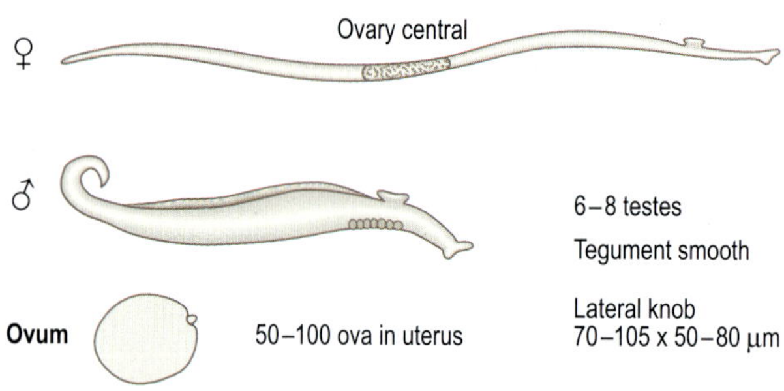

Host: *Oncomelania*

Distribution

S. haematobium: 78 million

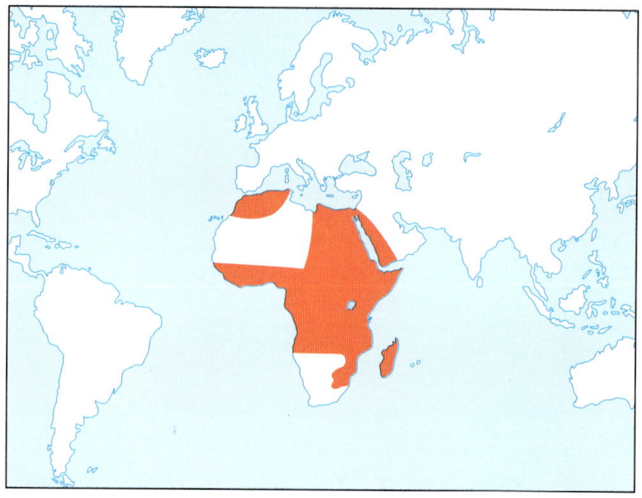

S. mansoni: 57 million

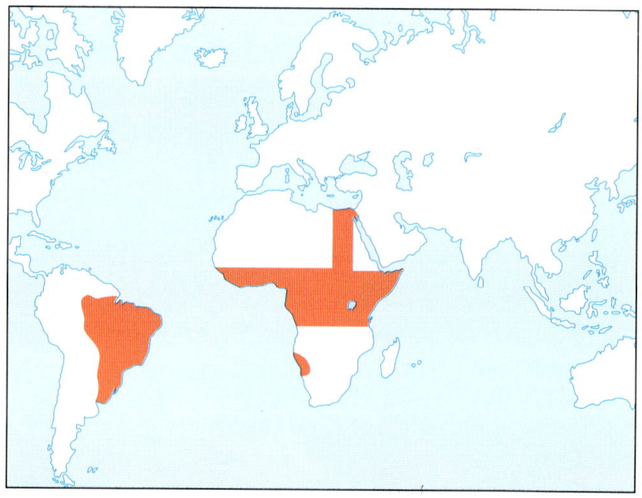

S. japonicum: 69 million

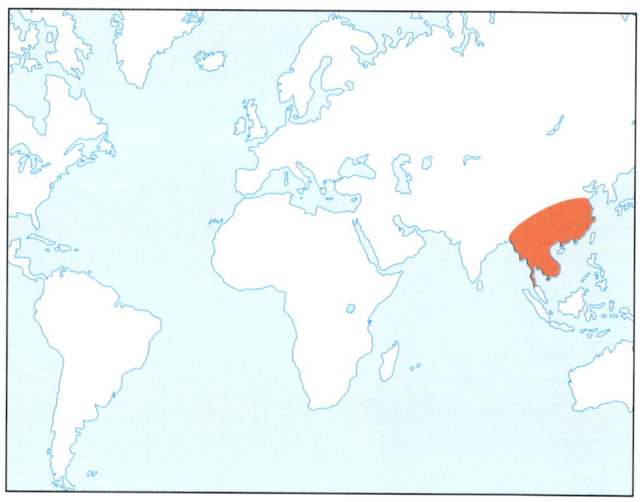

Schistosomiasis

Pathology

Penetration of the skin by cercariae (1)
Skin penetration may not be apparent. Human and some non-human *Schistosoma* species cause cercarial dermatitis (swimmer's itch). This manifests with papules, macules, vesicles and intense itching.

Migration and maturation of immature worms (2)
There are general toxic and allergic symptoms including urticaria with eosinophilia, fever, abdominal pain and tender hepatosplenomegaly. This is known as Katayama or snail fever.

Damage by eggs in tissue (3)
Resulting damage depends on the severity of the parasite load. An inflammatory granuloma forms with epithelial, giant, plasma and eosinophil cells and fibroblasts (Hoeppli reaction). There is subsequent fibrosis and calcification. Such damage may be local and/or ectopic.

Urinary schistosomiasis (4)
Caused by *S. haematobium*. Initial toxic and allergic symptoms are not marked, but the bladder and ureter are typically involved with hyperaemia, terminal haematuria, dysuria and frequency of micturition, papules, papillomata and ulceration. Hypertrophy of the bladder can lead to later contraction. There may be cystitis and calculus formation, with calcification and squamous cell carcinoma. Fistulae may develop. There can also be hydroureter and hydronephrosis. Ectopic lesions are less severe than in other species. Genital schistosomiasis may lead to lumpy semen, haematospermia or wart-like lesions on the vulva.

Intestinal schistosomiasis (5)
Caused by *S. mansoni*. There are marked initial toxic and allergic symptoms. The large intestine and rectum are typically involved with polyposis, papules, abscesses, ulcers, papillomata, fistulae and ova in faeces. The bladder is sometimes involved, with pathology as for urinary schistosomiasis as above. There can be ectopic lesions; the liver is frequently involved (receiving eggs via the portal vein with inflammatory reaction and fibrosis leading to periportal ('pipe-stem') fibrosis with portal hypertension, oesophageal varices, splenomegaly and ascites; there can also be lesions in the brain, spinal cord and lungs.

Oriental schistosomiasis (6)
Caused by *S. japonicum*. Initial toxic and allergic symptoms are marked and can lead to myocarditis and death. Intestinal lesions are similar to those with *S. mansoni* infection, and the small intestine is often involved. The liver is infected as in *S. mansoni*. Hepatic involvement occurs as for *S. mansoni*. The brain may also become involved.

General

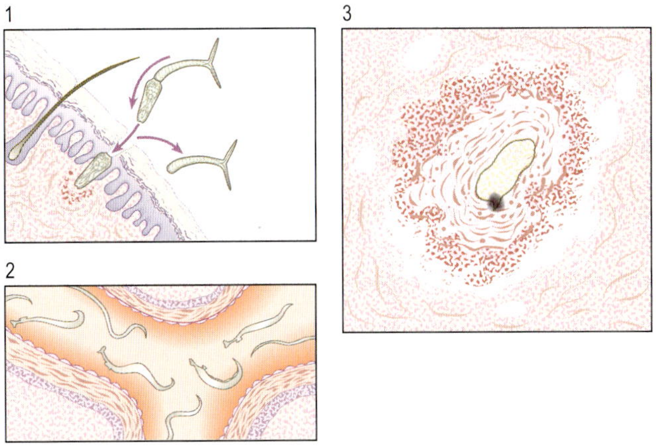

Particular

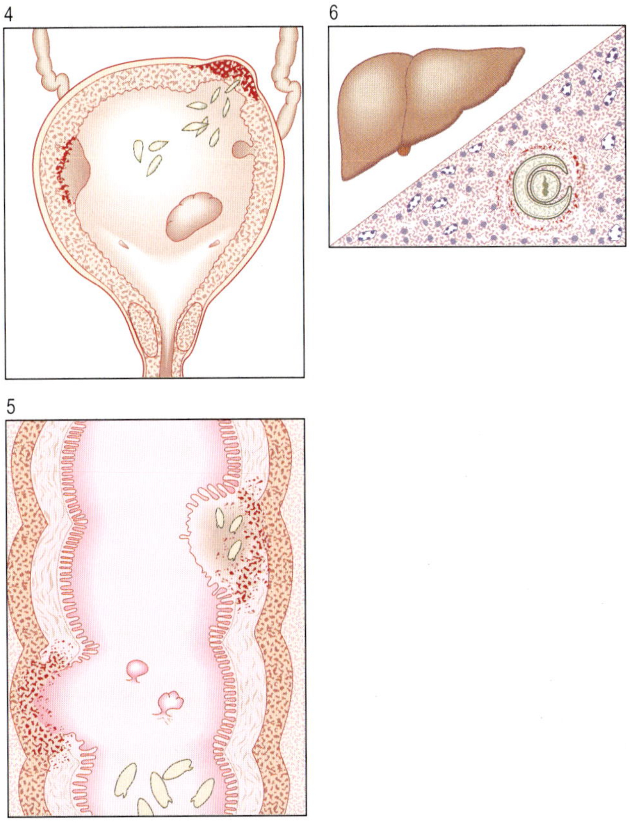

Laboratory diagnosis

Eosinophilia may be present.

Ova found in terminal urine by Nuclepore filtration or after centrifugation. Ova may also be found in semen. Ova may also be found in faeces directly or using formalin–ether concentration, rectal scrapings or biopsies.

Serology. ELISA tests (using soluble egg antigen) are useful 6–12 weeks post-exposure. In many chronic cases, the diagnosis will be made by serology alone.

Clonorchis sinensis, syn. *Opisthorchis sinensis* (Oriental liver fluke)

Life cycle

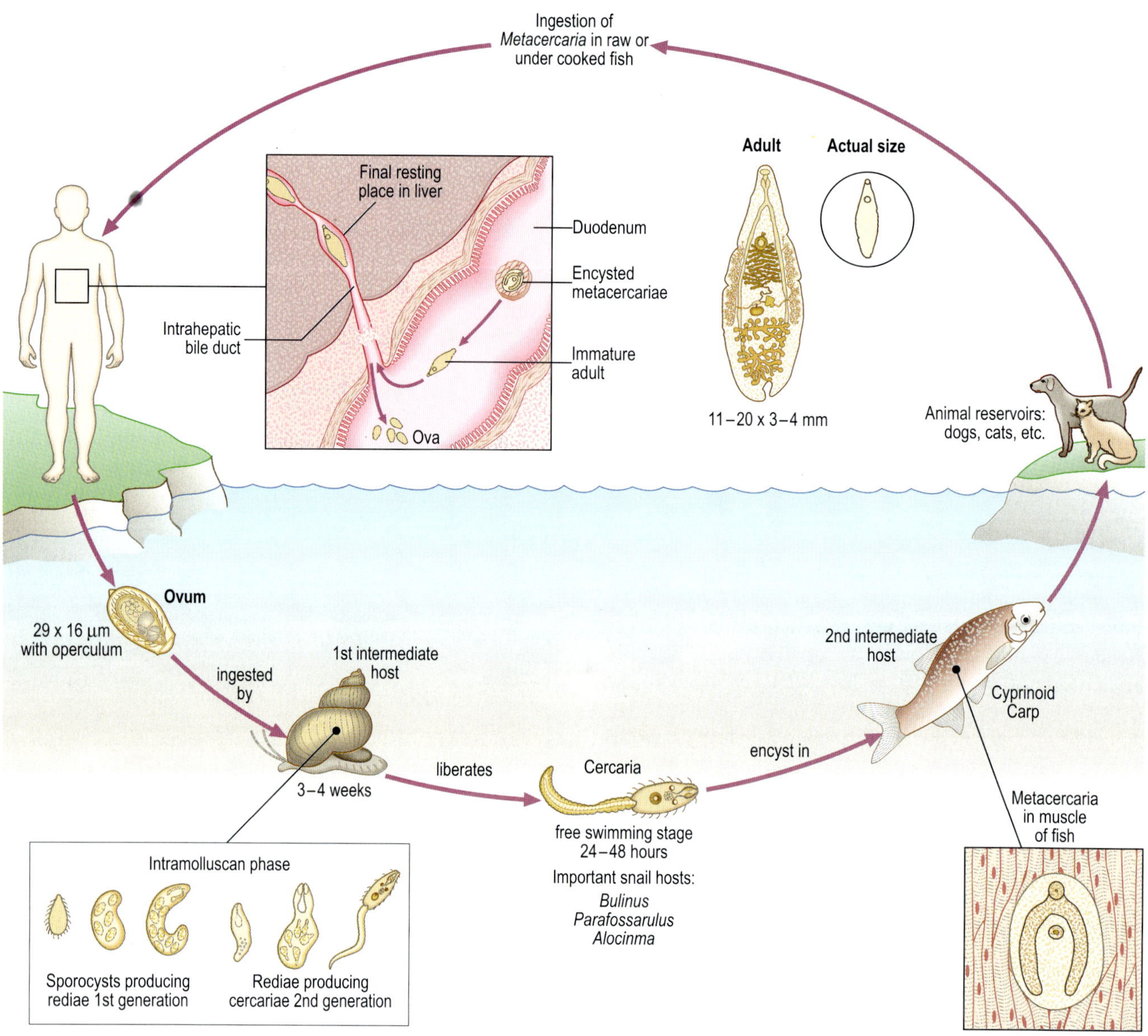

Pathology and Clinical features

Adult flukes inhabit the distal bile ducts with epithelial proliferation, surrounding inflammatory reaction and ascending cholangitis. Sometimes there is secondary bacterial infection with jaundice and septicaemia. There can also be eosinophilia. All this can lead to thick, dilated fibrous ducts with adenomata of epithelium, bile duct stenosis and cholangiocarcinoma. Many cases are asymptomatic. Acute infection may lead to tender hepatomegaly. Chronic infection can result in anorexia, low-grade fever, epigastric pain and tender hepatomegaly.

Laboratory diagnosis

Ova are found in faeces and in bile (via duodenal aspiration or 'string test').

Distribution

28 million infected worldwide.

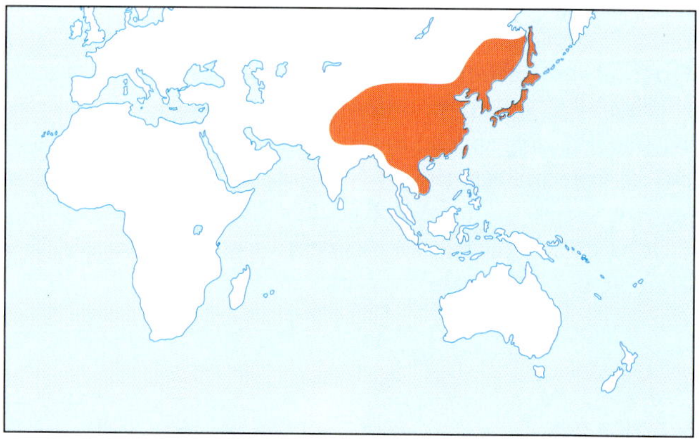

Clonorchis sinensis (continued)

Morphology

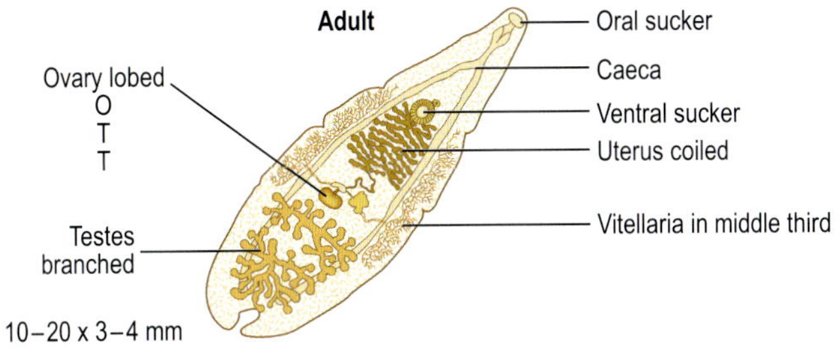

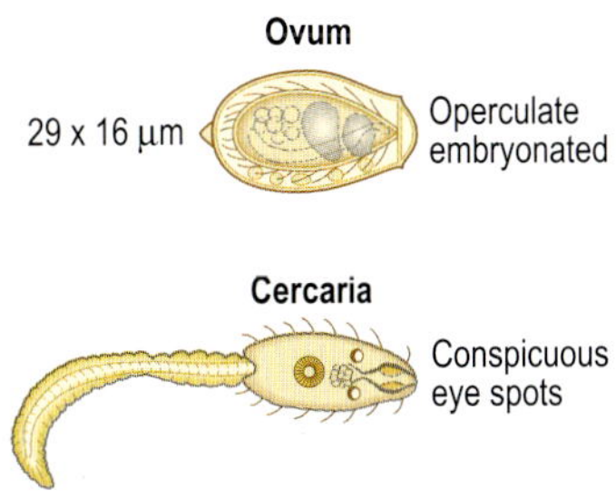

Opisthorchis felineus, Opisthorchis viverrini (cat liver fluke)

Morphology

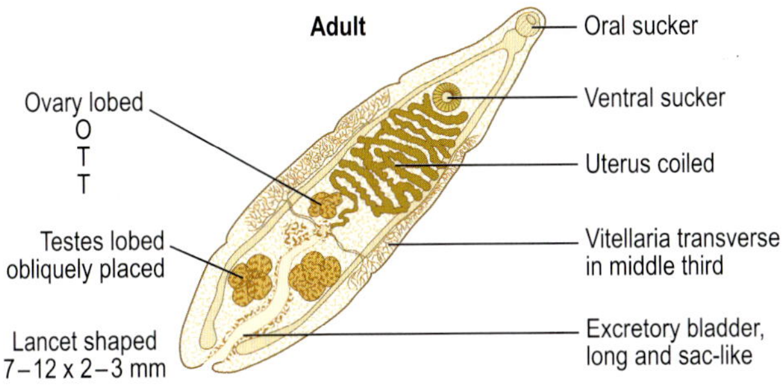

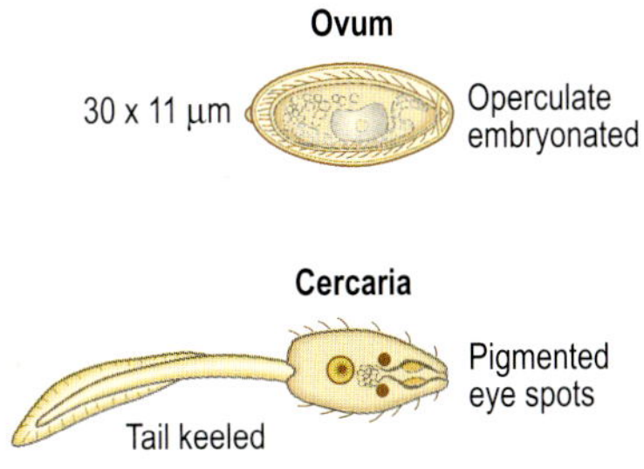

Life cycle

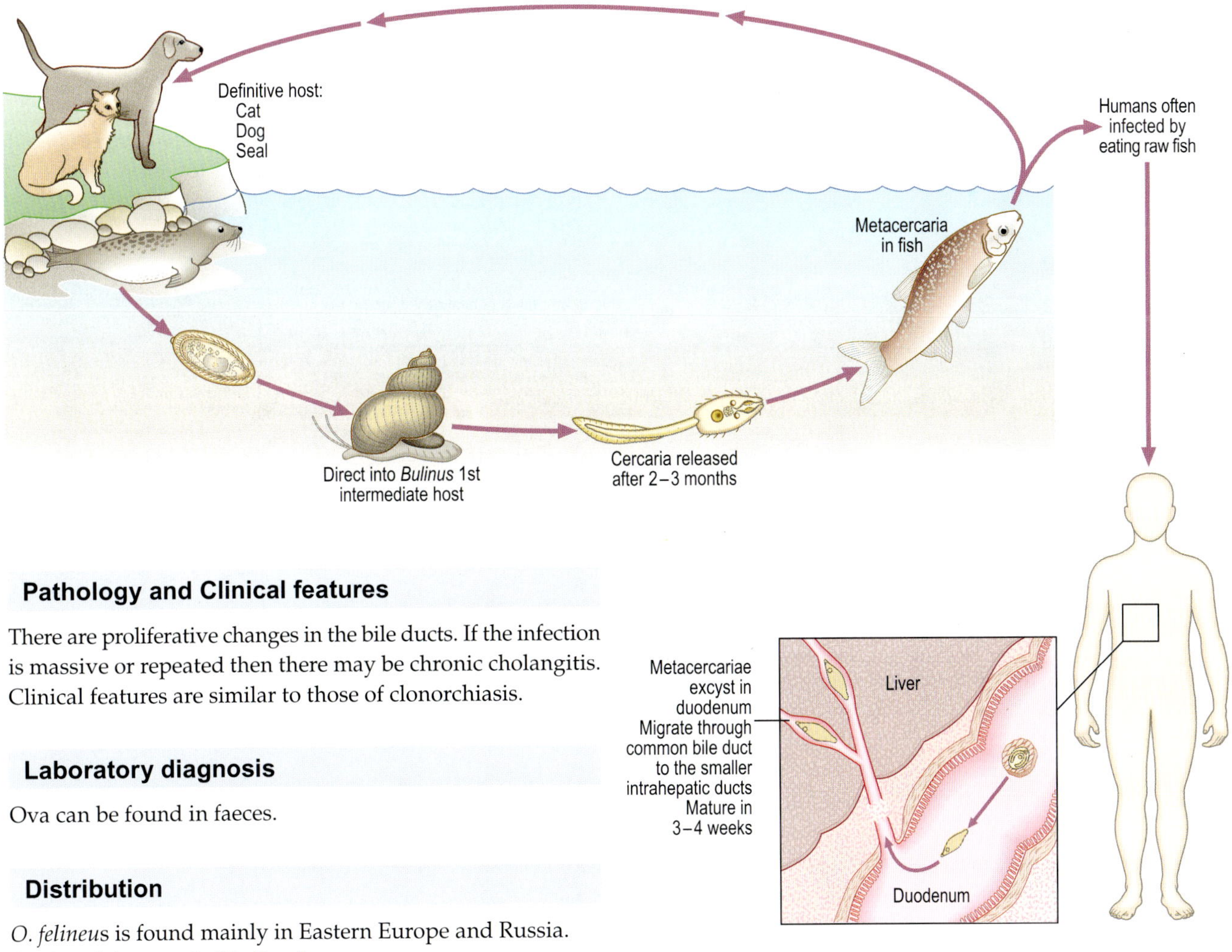

Pathology and Clinical features

There are proliferative changes in the bile ducts. If the infection is massive or repeated then there may be chronic cholangitis. Clinical features are similar to those of clonorchiasis.

Laboratory diagnosis

Ova can be found in faeces.

Distribution

O. felineus is found mainly in Eastern Europe and Russia. *O. viverrini* occurs in Thailand.

Fasciola hepatica (sheep liver fluke)

Life cycle

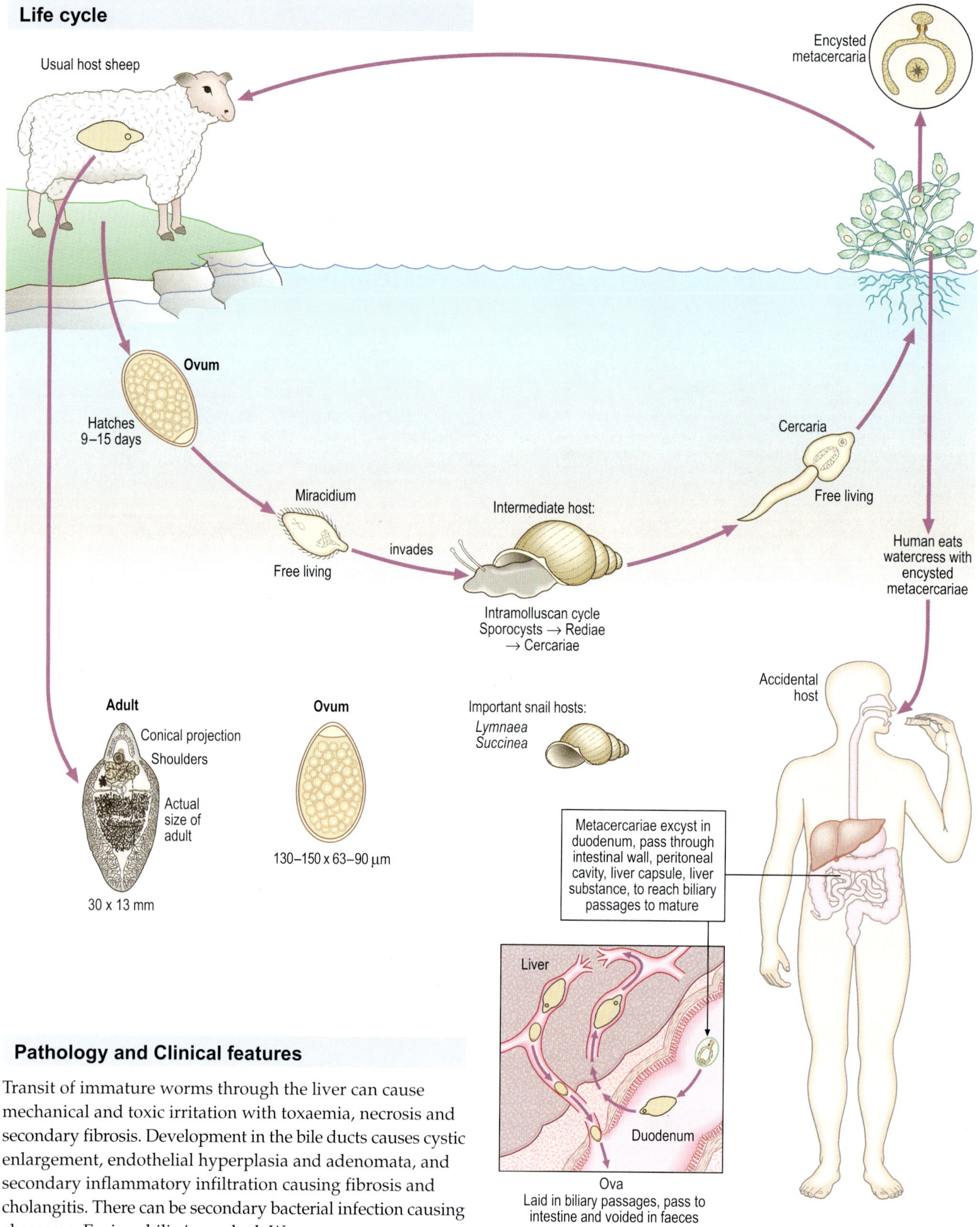

Pathology and Clinical features

Transit of immature worms through the liver can cause mechanical and toxic irritation with toxaemia, necrosis and secondary fibrosis. Development in the bile ducts causes cystic enlargement, endothelial hyperplasia and adenomata, and secondary inflammatory infiltration causing fibrosis and cholangitis. There can be secondary bacterial infection causing abscesses. Eosinophilia is marked. Worms can appear ectopically in lungs, brain, eyes, etc. with similar reactions. If raw sheep or goat's liver, infected by the adult fluke, is eaten there can be local irritation and pharyngeal infection (Halzoun).

Acute infection may present with fever, tender hepatomegaly, epigastric pain, anorexia and vomiting. Jaundice may occur. In chronic infection, there may be no symptoms or epigastric/right upper quadrant pain, hepatomegaly and vomiting.

Laboratory diagnosis

Ova are found in faeces. Serology (IFAT) is available.

Distribution

The fluke is found in all sheep-rearing countries. About 1 million people are infected worldwide.

Fasciolopsis buski

Life cycle

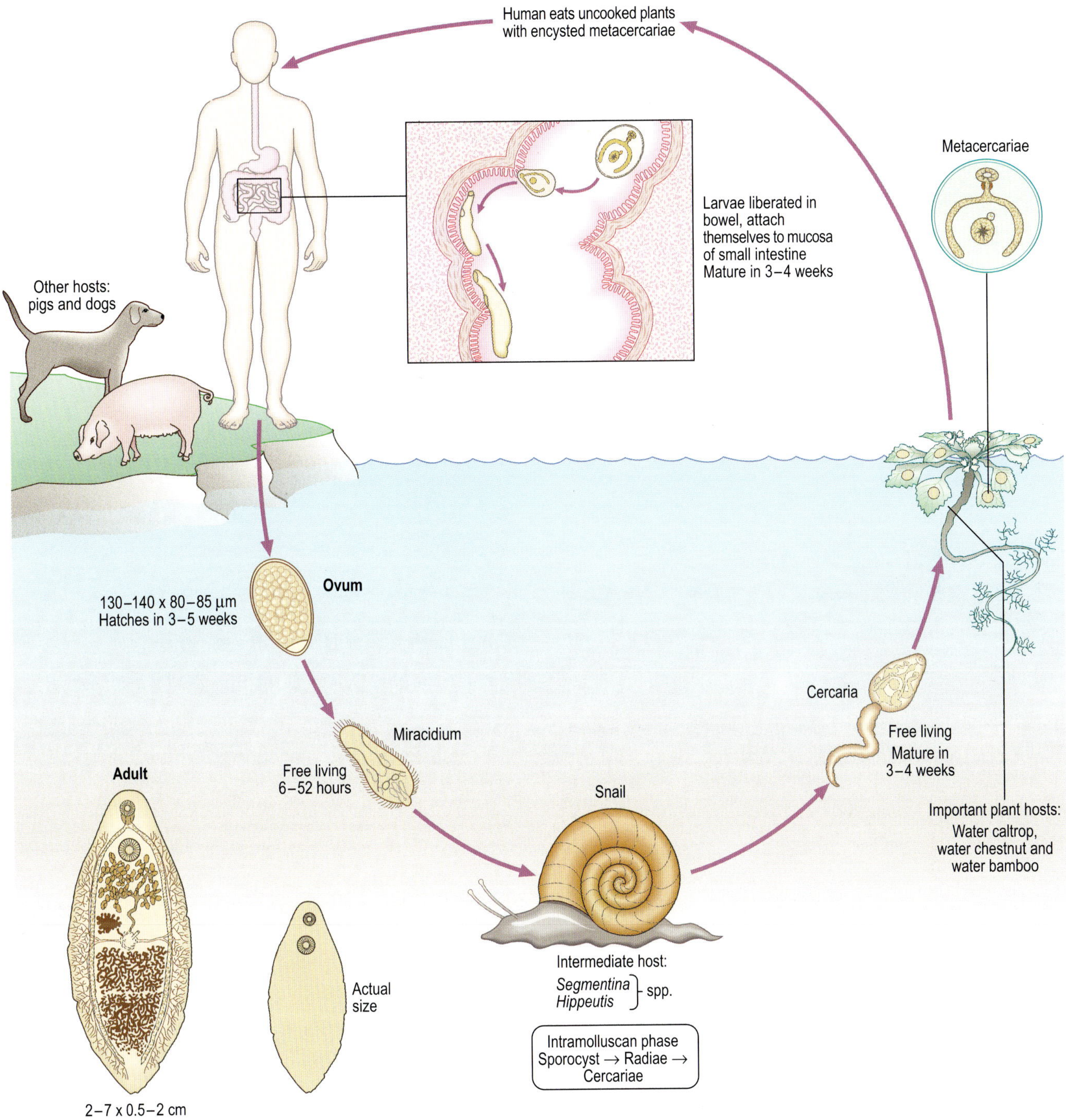

Pathology and Clinical features

There is localized inflammation at the site of attachment with haemorrhages and occasional abscesses. There is also eosinophilia. Lightly infected individuals may be asymptomatic. Diarrhoea, abdominal pain, anorexia, nausea and vomiting may occur.

Laboratory diagnosis

Ova, and sometimes adults, are found in faeces.

Distribution

15 million infected worldwide.

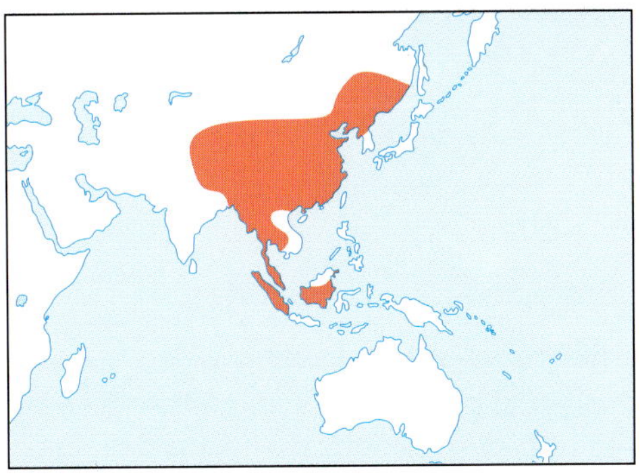

Heterophyes heterophyes

Morphology

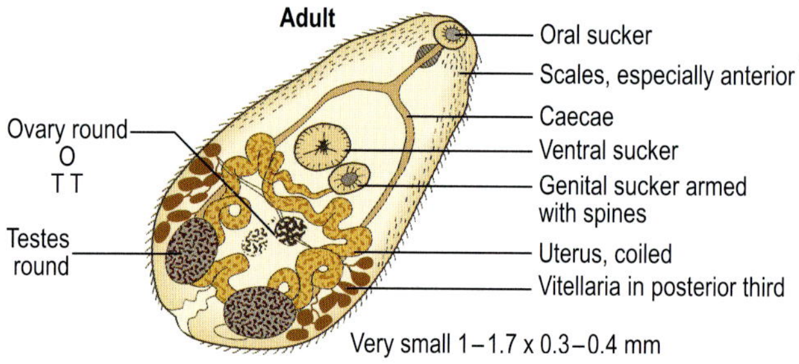

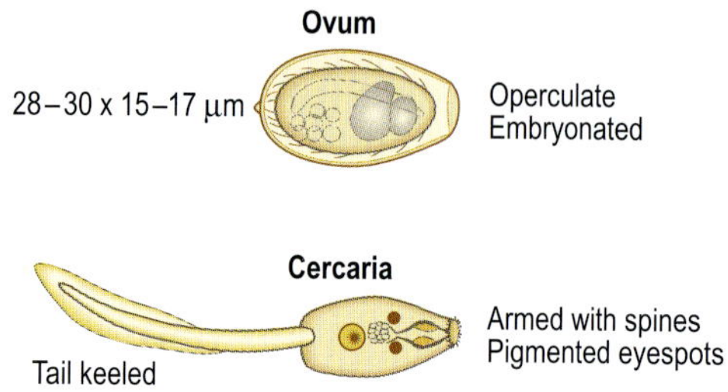

Life cycle

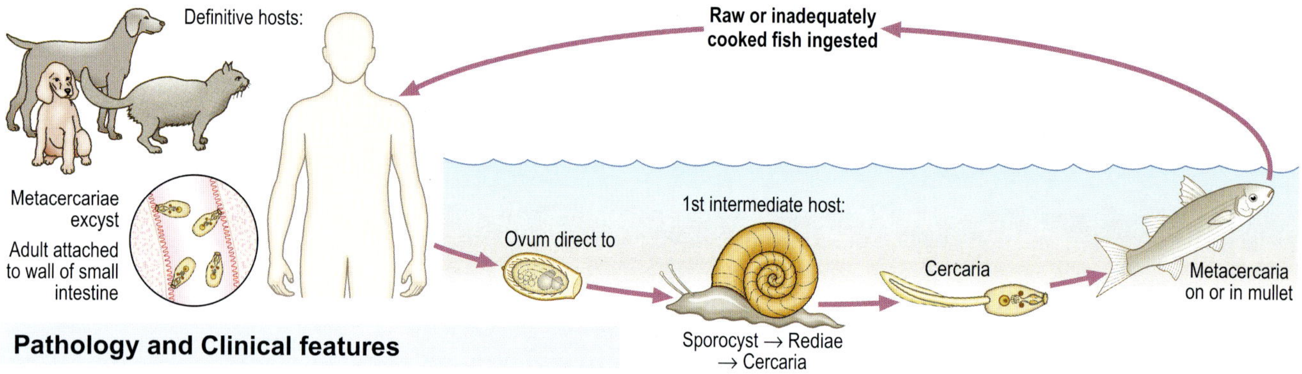

Pathology and Clinical features

There is a mild inflammatory reaction. Infected individuals may be asymptomatic or have abdominal pain, diarrhoea, anorexia and nausea. Ectopic ova have been found in the heart and brain.

Distribution and laboratory diagnosis

Found in the Nile Delta and the Far East. Ova in faeces.

Metagonimus yokogawai

Morphology

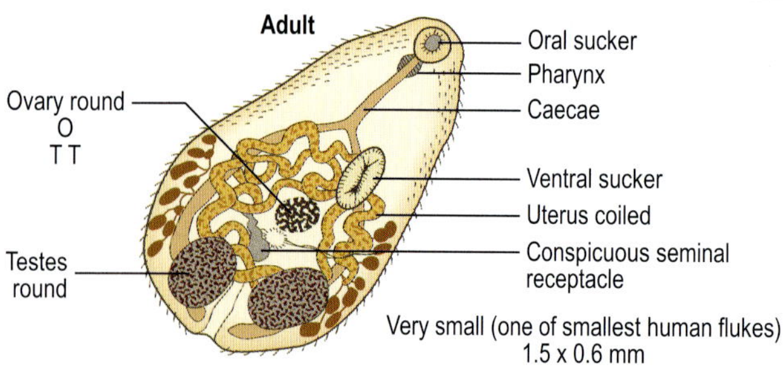

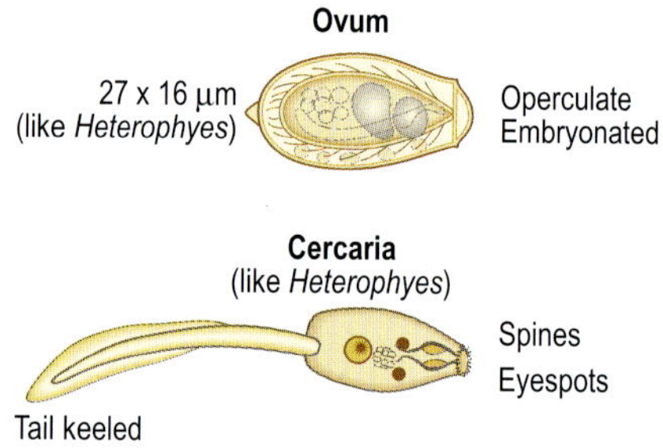

Life cycle

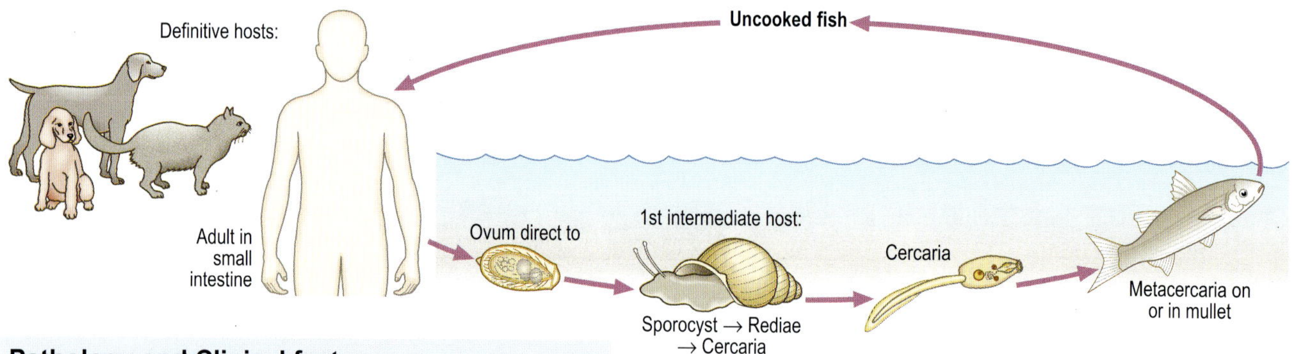

Pathology and Clinical features

Causes mild inflammatory reaction in the intestine. Occasionally ectopic ova can cause granulomata in other organs of the body, especially the liver and brain.

Distribution

Prevalent in the Far East.

Paragonimus westermani (lung fluke)

Life cycle

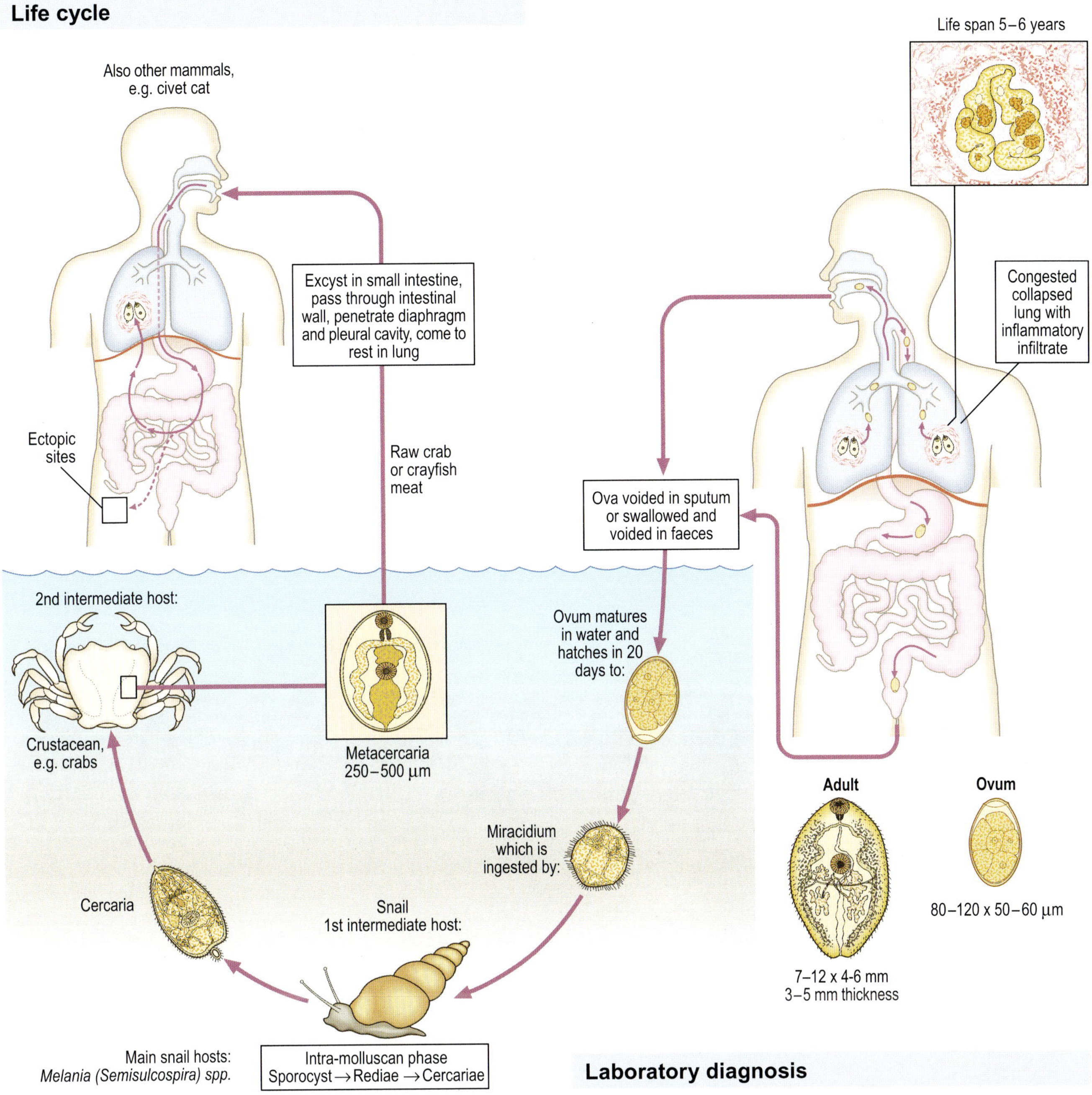

Pathology and Clinical features

The initial invasion has little pathological effect on the host. On localization in the lungs, there is tissue reaction leading to formation of a fibrous tissue capsule (of a slate blue colour) containing worms (generally in pairs), ova and inflammatory infiltrate. The capsule is connected with the respiratory passages. Secondary complications of these lung cysts include bronchiectasis, abscess formation and haemoptysis. Localization in other sites can cause cysts in any part of the body (for example the brain, causing epilepsy). Eosinophilia is a general manifestation. Chronic infection may be asymptomatic. Cough, brown gelatinous sputum, chest discomfort, shortness of breath and pleuritic chest pain may occur.

Laboratory diagnosis

Ova are found in sputum after KOH digestion or faeces after formalin-ether concentration. Serological tests, when available, are CF or ELISA (using extract of adult flukes as antigen) or gel diffusion. Chest X-ray or CT can also be used.

Distribution

5 million infected worldwide.

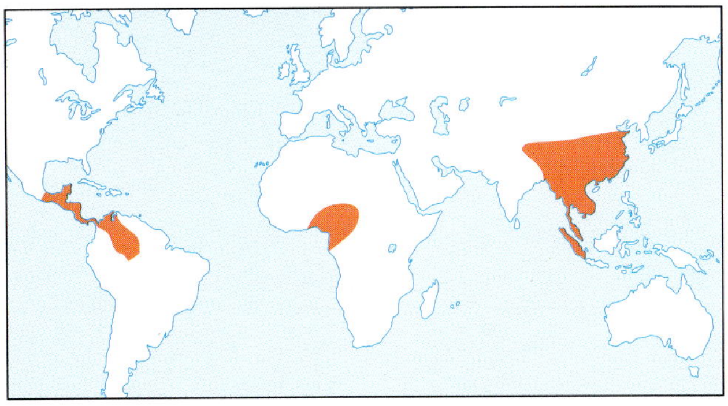

Common helminth ova

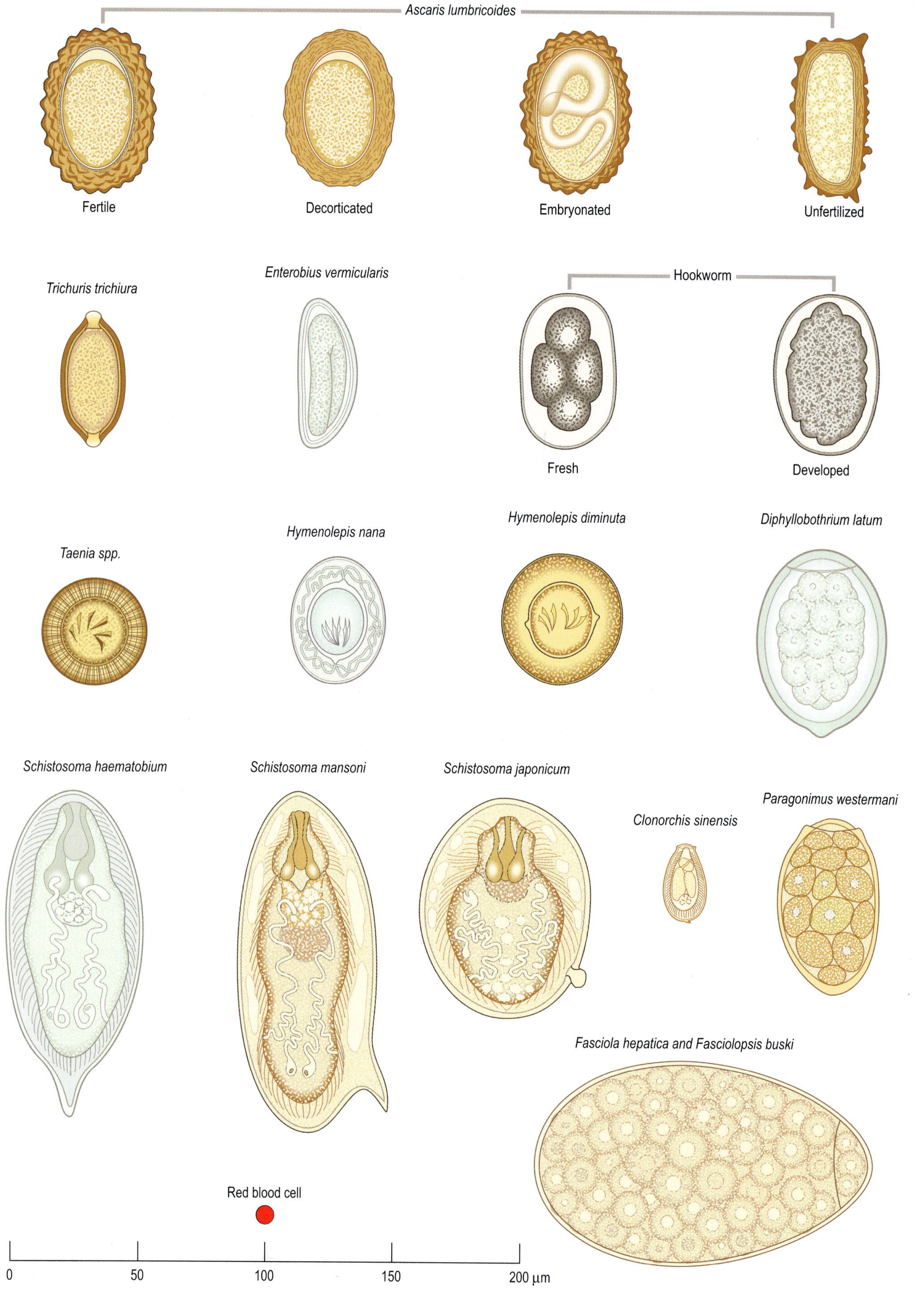

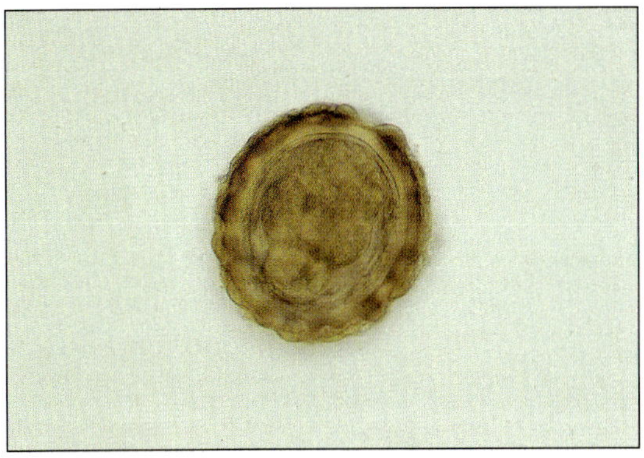

(a) *Ascaris* ovum

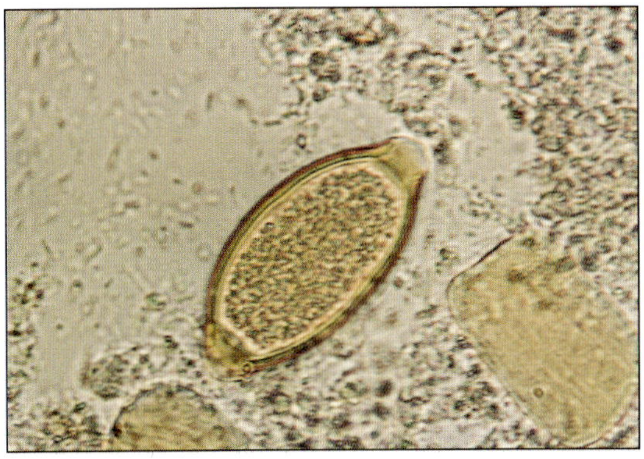

(b) *Trichuris* ovum

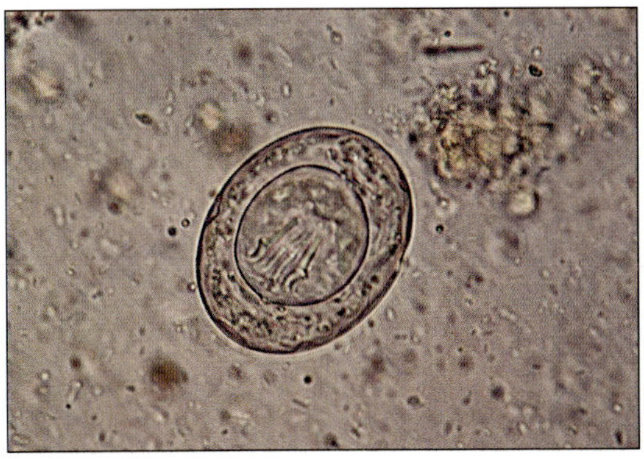

(c) *Hymenolepis nana* ovum

(d) *Schistosoma mansoni* ovum

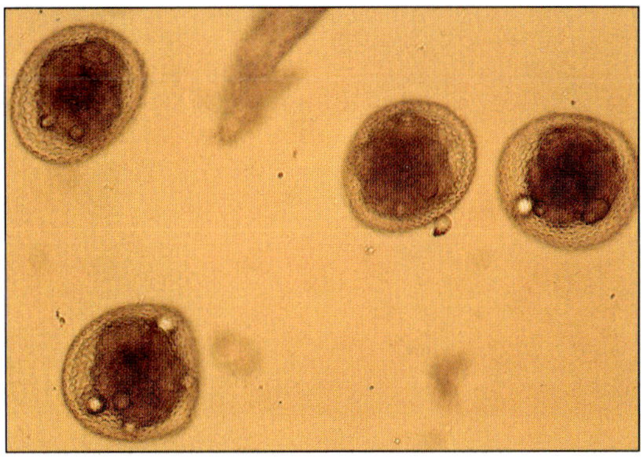

(e) *Toxocara canis* ova

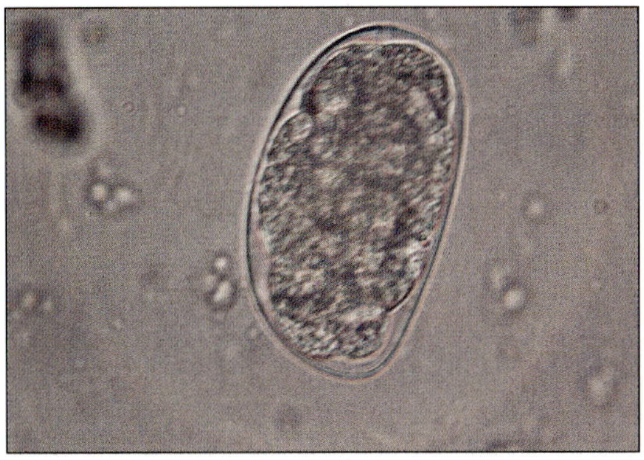

(f) Hookworm (*Ancylostoma*) ovum

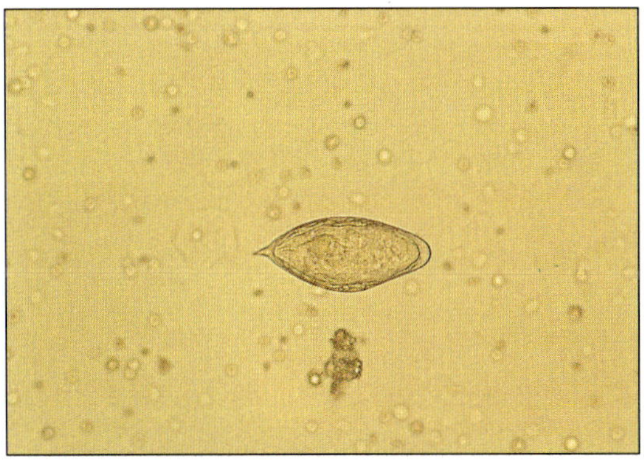

(g) *Schistosoma haematobium* ovum

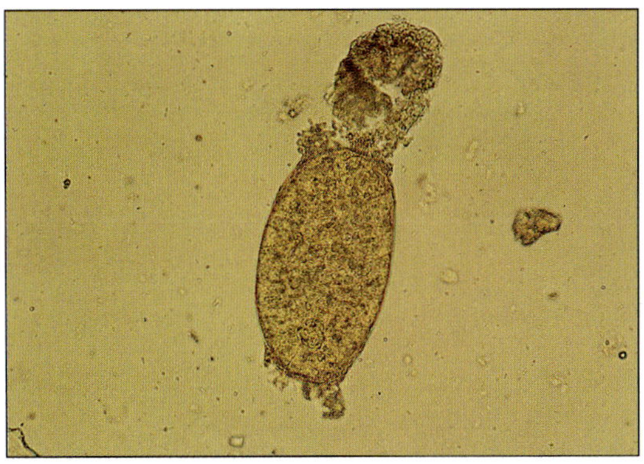

(h) *Fasciola hepatica* ovum

Protozoology

An outline classification of the parasitic protozoa of humans

Empire	Kingdom	Phylum	Class	Order	Genus
Eukaryota	Archezoa Haeckel 1894	Metamonada	Trepomonadea	Diplomonadida Enteromonadida	*Giardia* *Enteromonas*
			Retortamonadea	Retortamonadida	*Chilomastix* *Retortamonas*
		Microspora	Microsporea	Microsporida	*Encephalitozoon* *Enterocytozoon* *Nosema* *Septata* *Trachipleistophora*
	Protozoa Goldfuss 1818	Percolozoa	Heterolobosea	Schizopyrenida	*Naegleria*
		Parabasala	Trichomonadea	Trichomonadida	*Dientamoeba* *Trichomonas*
		Euglenozoa	Kinetoplastidea	Trypanosomatida	*Leishmania* *Trypanosoma*
		Ciliophora	Litostomatea	Vestibuliferida	*Balantidium*
		Apicomplexa (Sporozoa)	Coccidea	Eimeriida	*Cryptosporidium* *Cyclospora* *Isospora* *Sarcocystis* *Toxoplasma*
			Haematozoea	Haemosporida Piroplasmida	*Plasmodium* *Babesia*
		Rhizopoda	Lobosea	Acanthopodida	*Acanthamoeba* *Balamuthia*
			Entamoebidea	Euamoebida	*Endolimax* *Entamoeba* *Iodamoeba*

Intestinal protozoa

Coccidia

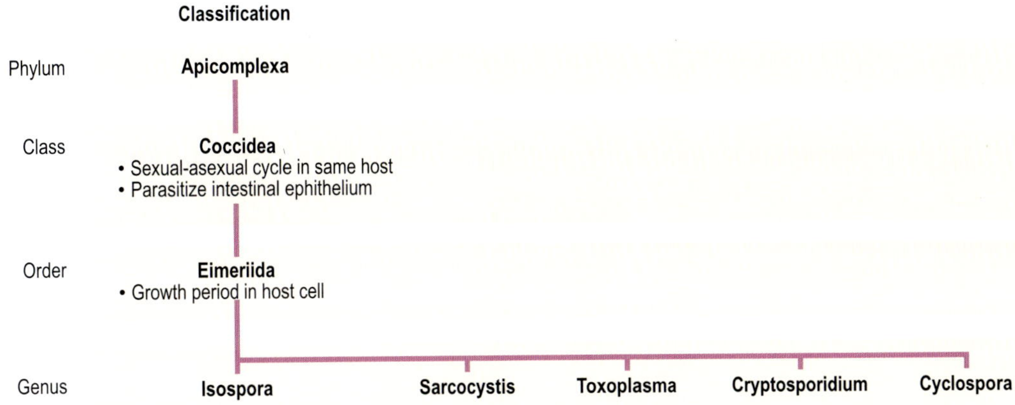

Isospora belli (causing coccidiosis in humans)

Life cycle

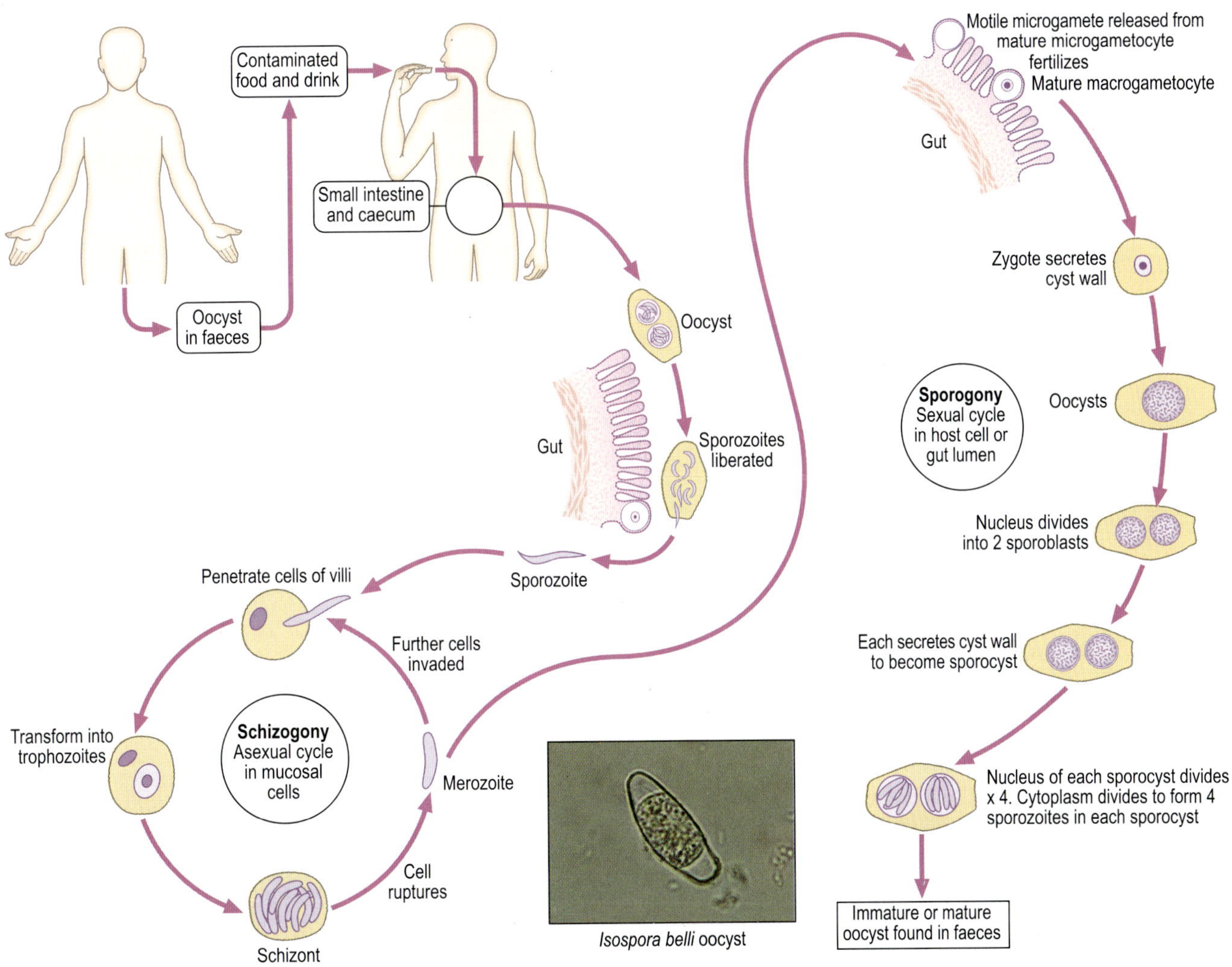

Isospora belli oocyst

Pathology and Clinical features

Small bowel mucosal atrophy. Watery diarrhoea or steatorrhoea, weight loss and sometimes cholecystitis occur in AIDS.

Laboratory diagnosis

Oocysts are seen in formalin–ether concentration of faeces or modified Ziehl-Neelsen or auramine-stained faecal smears. Intraepithelial parasites may be seen in small bowel biopsies.

Cryptosporidium parvum

Life cycle

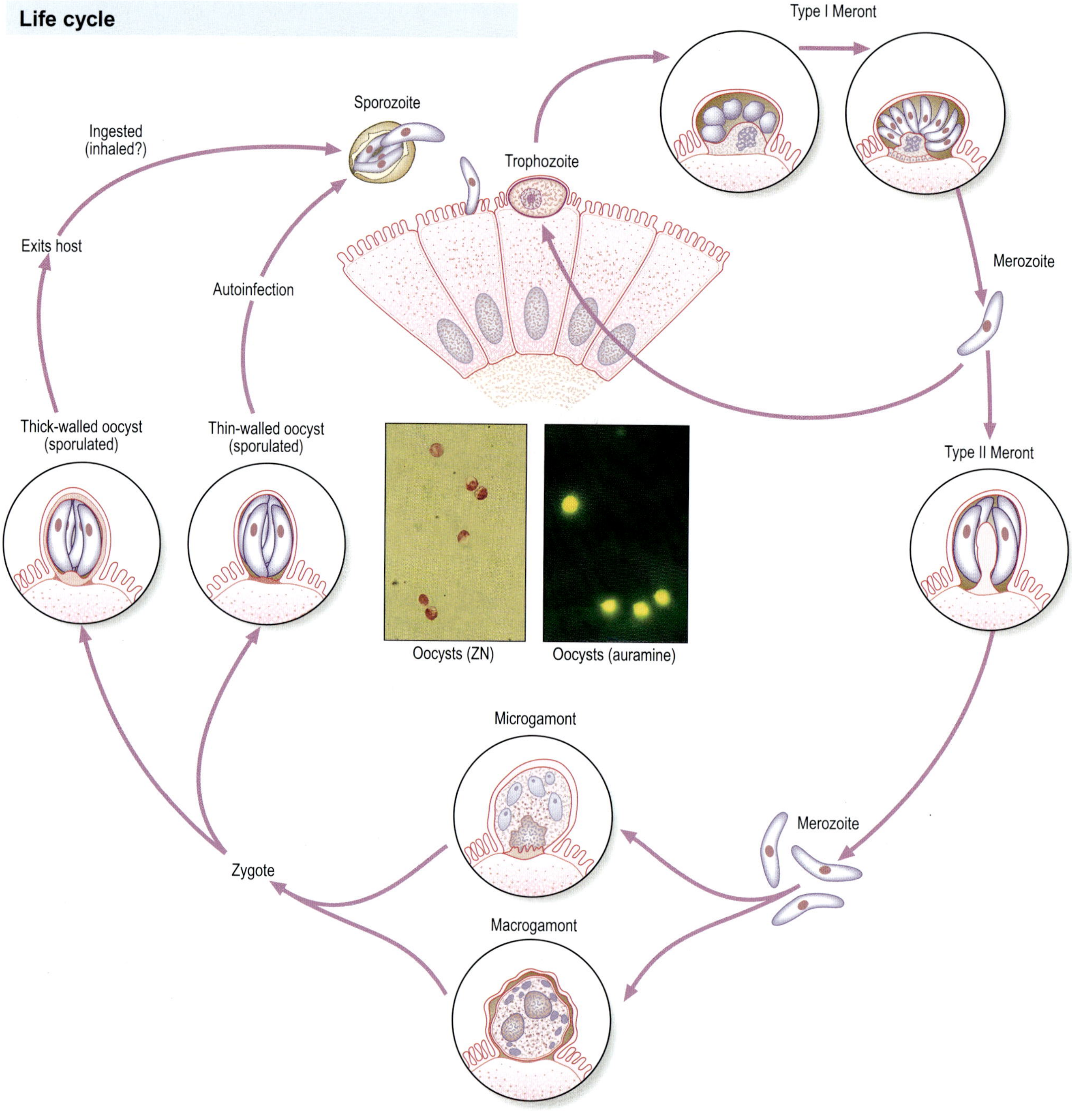

Pathology and Clinical features

In the immunocompetent, there is short-term enteropathy with self-limiting diarrhoea. In the immunocompromised, for example a patient with AIDS or a child with severe combined immunodeficiency, there is chronic diarrhoea with malabsorption and weight loss. Extraintestinal infection of the respiratory tract, biliary tract and pancreas may occur.

Distribution

Cryptosporidia have a cosmopolitan distribution. Human and farm animal strains exist; both can cause human disease. Human infection is usually waterborne.

Laboratory diagnosis

Oocysts (4–5 μm in diameter) are found in faeces, using modified Ziehl-Neelsen stain, auramine or specific FITC labelled monoclonal antibody staining. They can also be found in faecal concentrates, duodenal aspirates and duodenal biopsies. The oocysts are very small (5 μm in diameter) and round. Parasites may also be seen in small bowel biopsies. Sucrose floatation is an alternative to formalin–ether concentration.

Cyclospora cayetanensis

Life cycle

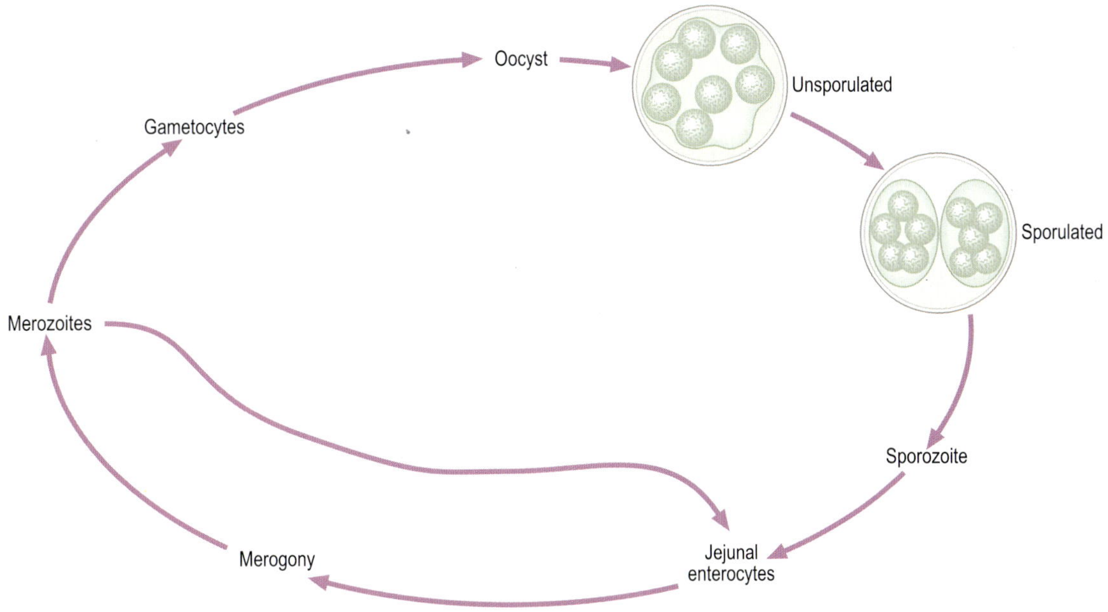

Cyclospora

Oocysts are 8–10 µm in diameter with a central morula of refractile spheres when unsporulated. These mature into a final division of 2 sporocysts.

Pathology and Clinical features

Acute onset of diarrhoea, followed by steatorrhoea. Colicky abdominal pain and malaise. Partial villous atrophy may be seen.

Laboratory diagnosis

Oocysts are seen in faeces unsporulated when first passed. Diagnosis is either by formalin–ether concentration, modified Ziehl-Neelsen stain or by autofluorescence.

Distribution

Widespread, probably worldwide.

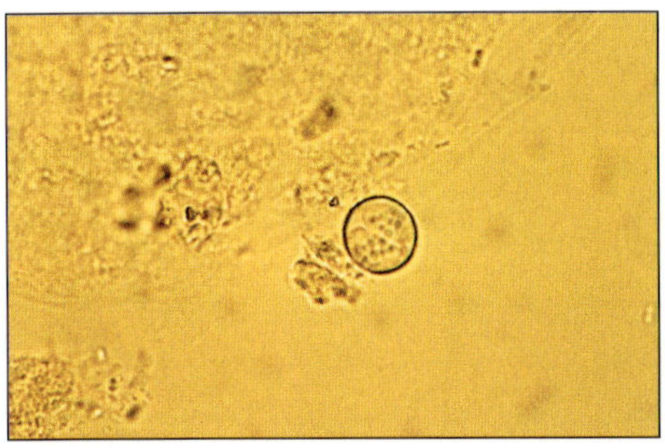

Sporulated

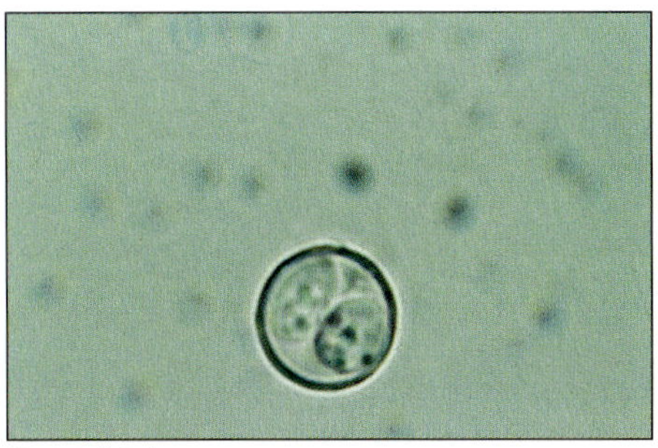

Unsporulated

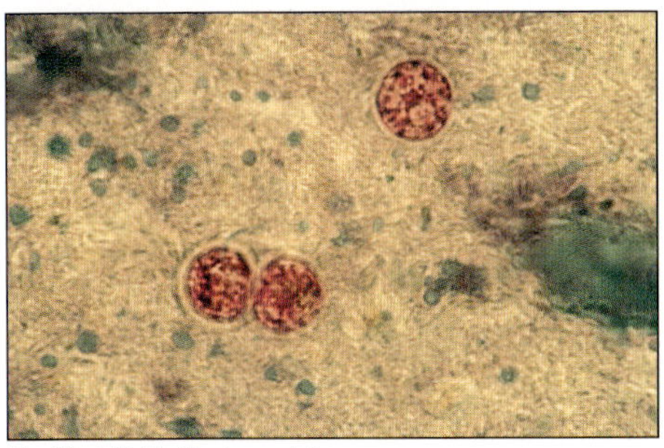

Modified Ziehl–Neelsen stain

Autofluorescence

Sarcocystis hominis

Probable life cycle

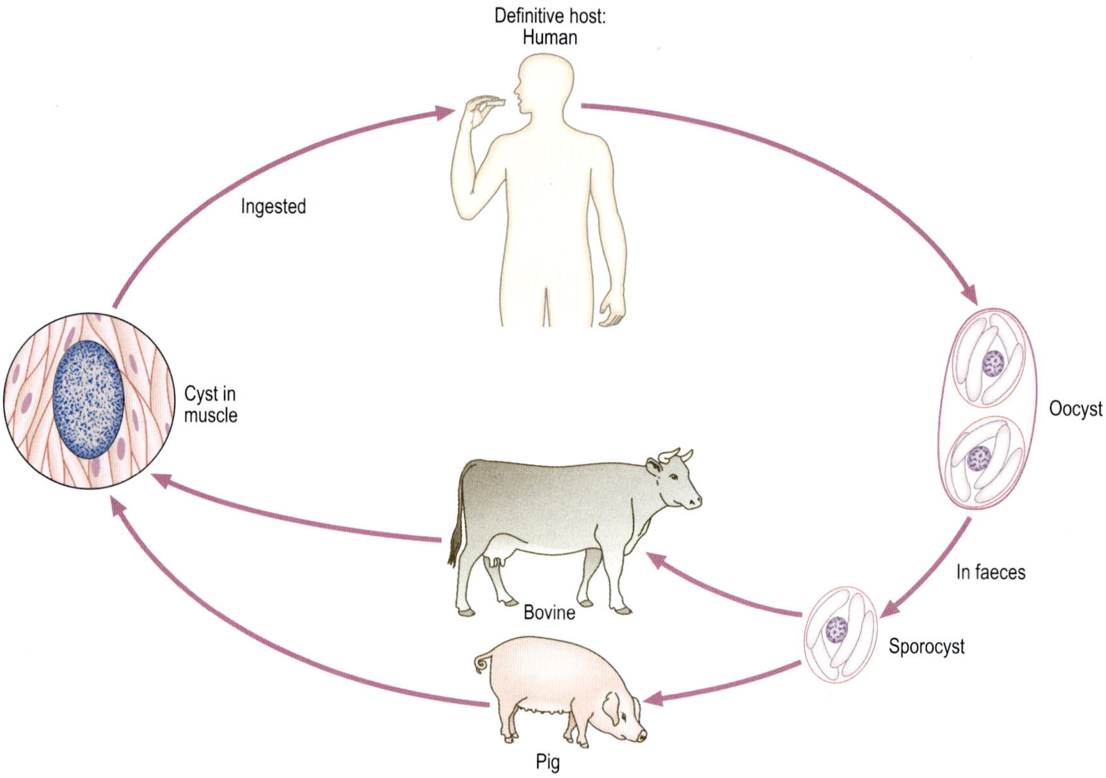

Occasionally humans can act as intermediate hosts for *Sarcocystis* of other animals.

Morphology

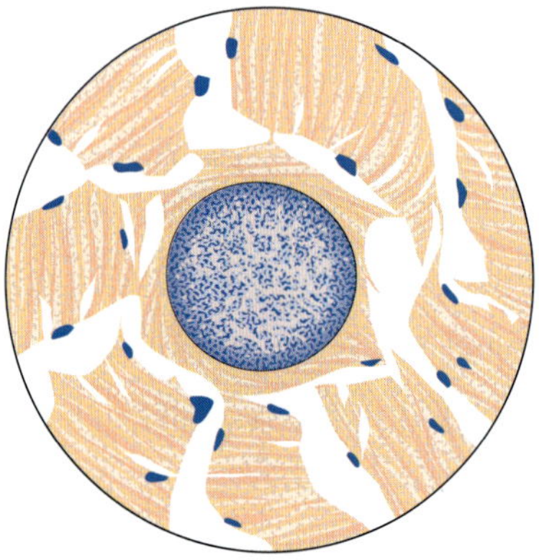

Cyst in human muscle
Miescher's tube x 100

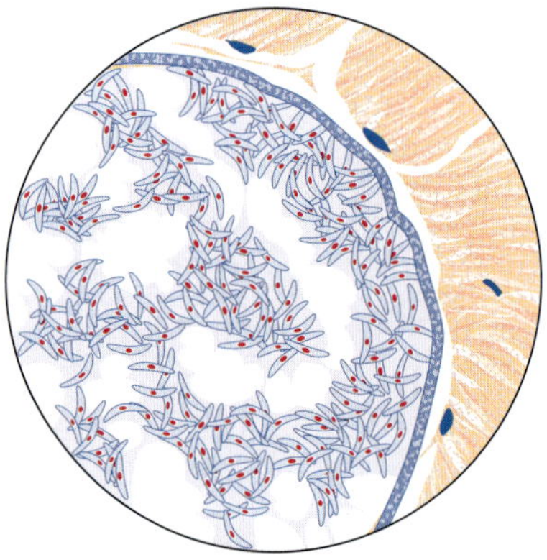

Enlarged portion of Miescher's tube showing Rainey's corpuscles (each 12–16 x 4–9 μm): from a human case

Pathology and Clinical features

The intestinal stages produce diarrhoea and abdominal pain. The clinical significance of muscle cysts is unknown.

Laboratory diagnosis

Oocysts or free sporocysts are found in faeces. Histological examination of biopsy specimens may show the sexual stages in the intestinal epithelium.

Histology is the only way to diagnose the presence of sarcocysts, although these are almost invariably incidental findings.

Microsporidia–general characteristics

All are obligate intracellular parasites. The vast majority of species are in invertebrates, especially insects, lower vertebrates and fish. Only a few have been reported from warm-blooded vertebrates.

They are considered to be primitive organisms. Their evolutionary history has been predicted from their prokaryote-like ribosomal characteristics — the absence of a separate 5.8S rRNA and the nucleotide sequence of the small subunit (16S) rRNA. They have no mitochondria. The infective stages are highly-resistant spores. These are very uniform in size for a given species.

When spores are ingested by a new host, the cells are penetrated by means of an apparatus known as the polar tube. When this is fully extended, the sporoplasm passes through the tube, to be inoculated into the cytoplasm of the host cell.

Following infection, there follows a phase of multiplication by binary or multiple fission (merogony). The transition to the spore-producing stage (sporogony) is heralded by the secretion of an electron dense surface coat — this will form the future exospore layer of the spore wall. The primary sporogonic cells are **sporonts**, which divide into **sporoblasts**, which mature into **spores**, which are released when the host cell ruptures.

Common species of microsporidia reported from humans. Most are AIDS associated.

Species	Localization	Pathogenesis
Encephalitozoon cuniculi	Generalized, brain, etc.	Convulsions, etc.
Encephalitozoon hellem	Corneal epithelia	Keratopathy
Enterocytozoon bieneusi	Enterocytes–gut	Diarrhoea
Encephalitozoon (Septata) intestinalis	Enterocytes–gut	Diarrhoea
Nosema connori	Generalized	Multi-organ
Nosema corneum	Corneal stroma	Keratitis
Microsporidium africanum	Corneal stroma	Keratitis
Pleistophora sp.	Muscle fibres	Myositis

Infections of the gastrointestinal tract and urinary system can be detected by the presence of spores in faeces or urine. Spores from these sites can be visualized by staining them with the modified trichrome stain.

The spores of microsporidia are very small–1 x 0.5 μm (See below)

Enterocytozoon bieneusi spores

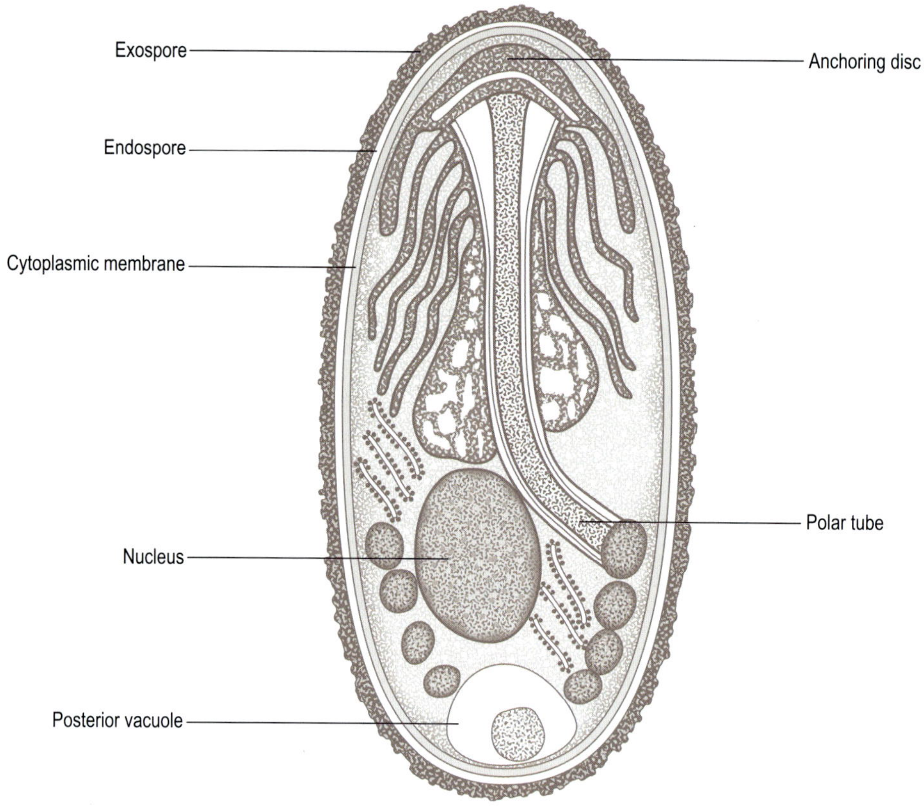

Microsporidian spore

Life cycle

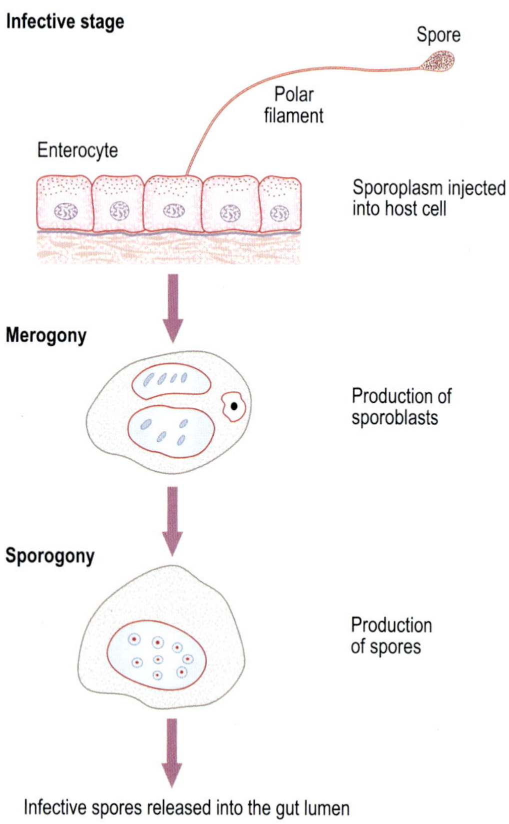

Laboratory diagnosis

Alternative staining methods for microsporidial spores in stool samples are modified trichrome stain and uvitex 2B or calcofluor fluorescence.

Entamoeba
- Generally one nucleus in trophozoite
- Small karyosome at or near centre
- Nuclear membrane lined with chromatin granules
- Forms cysts

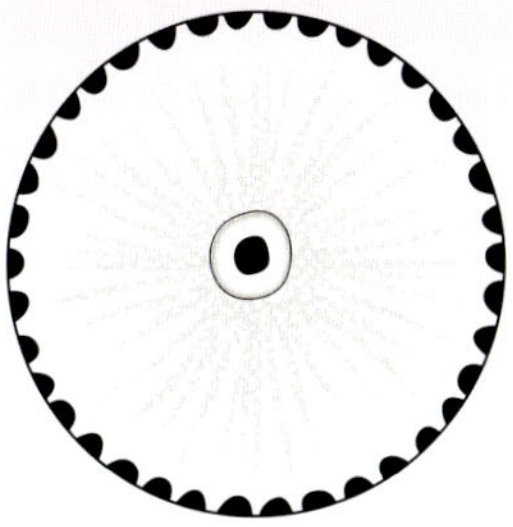

Endolimax
- Generally one nucleus in trophozoite
- Large irregular karyosome attached to nuclear membrane
- No peripheral chromatin
- Forms cysts

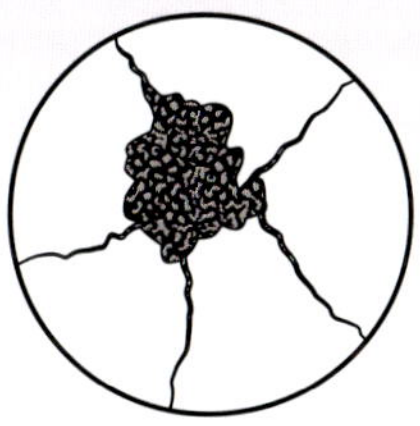

Iodamoeba
- Generally one nucleus in trophozoite
- Large karyosome surrounded by achromatic granules
- No peripheral chromatin
- Forms cysts

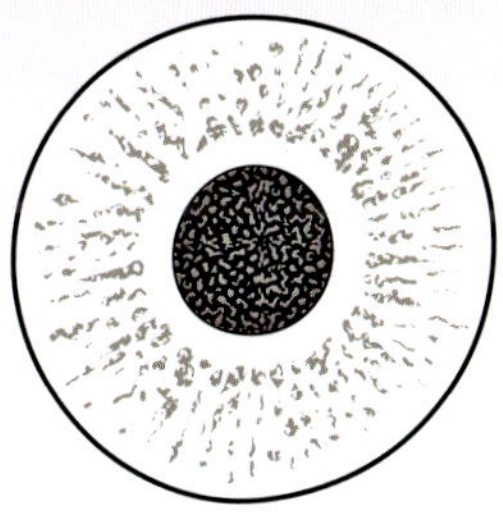

Dientamoeba
- Minute
- Generally binucleate
- Central particulate karyosome
- No peripheral chromatin
- No cystic stage

Species

Entamoeba histolytica/dispar
Entamoeba coli
Entamoeba hartmanni
Entamoeba polecki

Endolimax nana

Iodamoeba bütschlii

Dientamoeba fragilis

Entamoeba histolytica (causing amoebiasis)

Life cycle

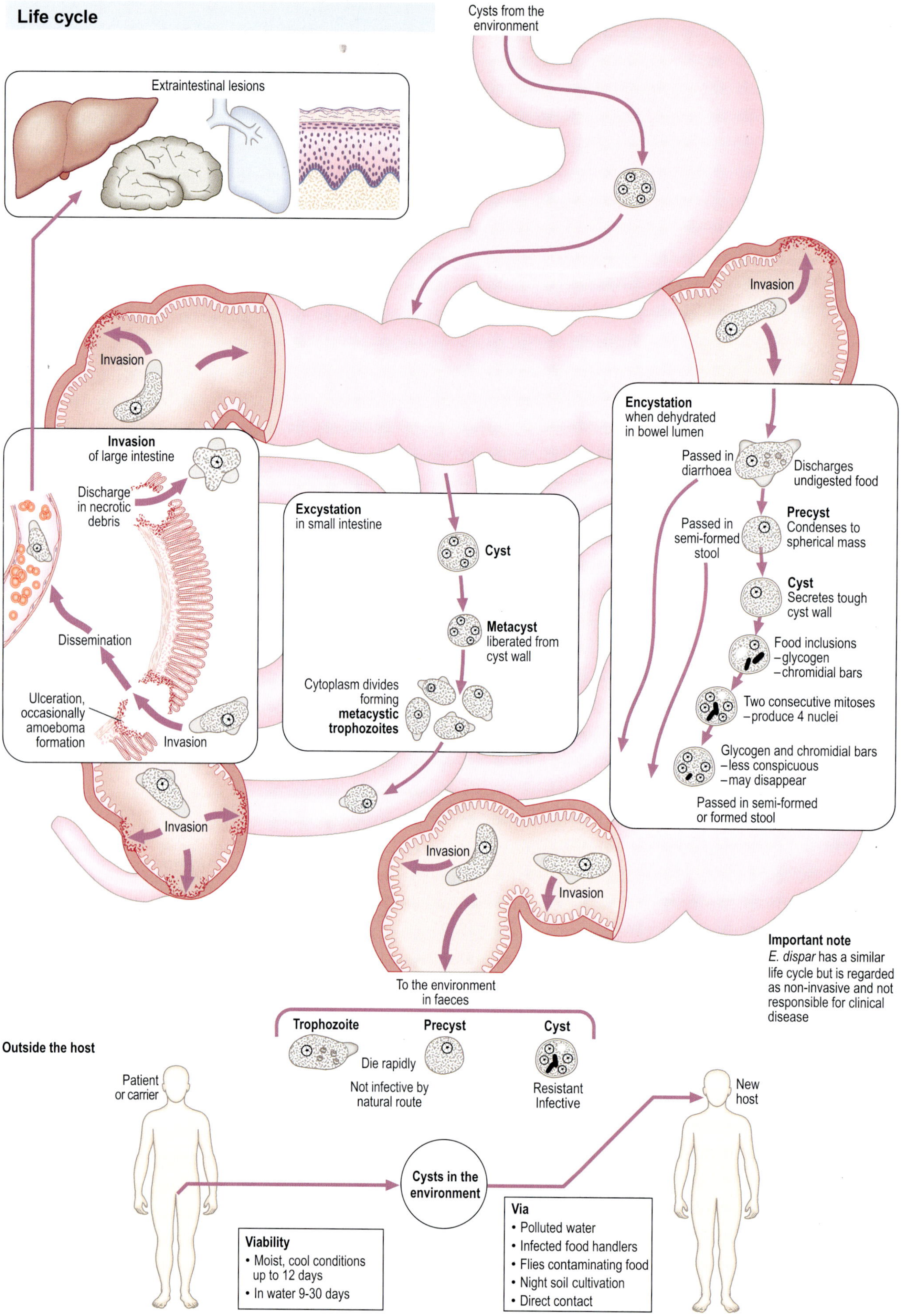

Morphology

General–nomenclature

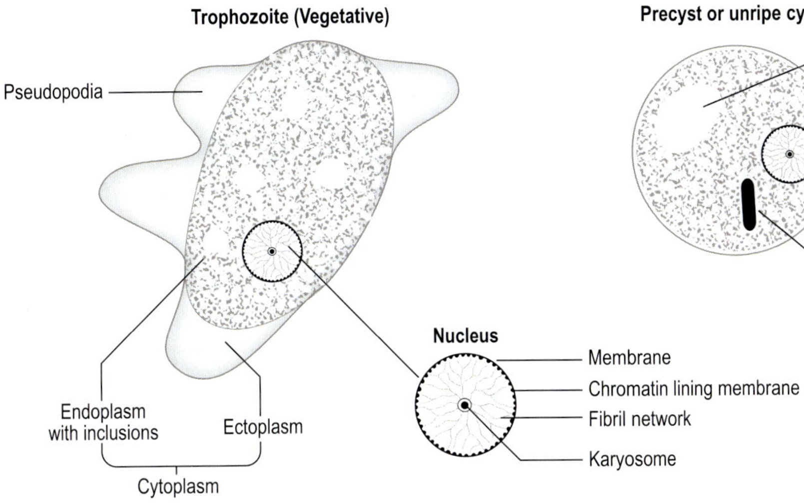

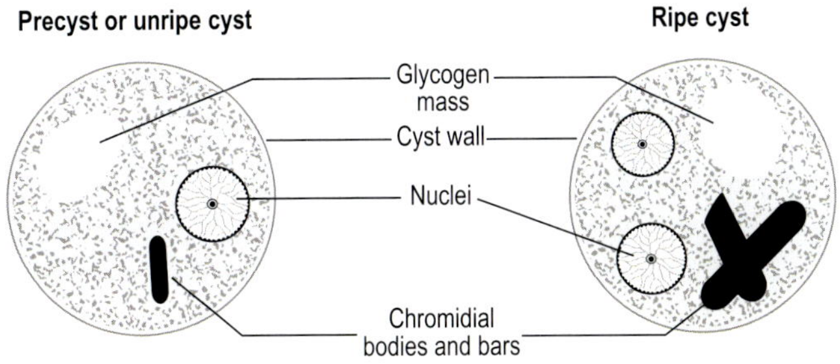

Important note
E. dispar is morphologically identical to *E. histolytica* but the trophozoites are not haematophagous

Particular - Includes differentiation from *Entamoeba coli*, an intestinal commensal.

Unstained preparations

Trophozoite

E. histolytica				*E. coli*
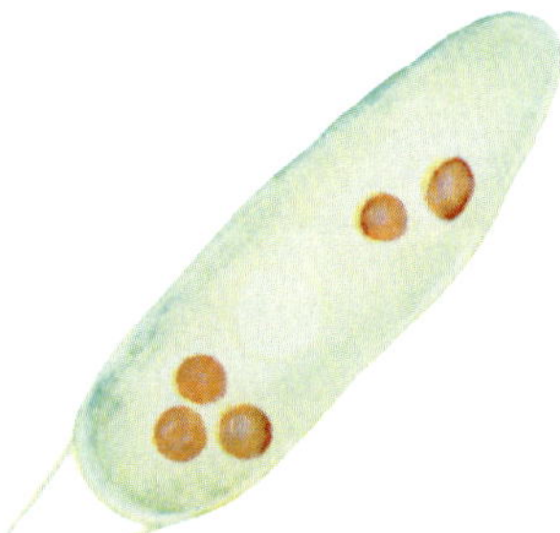 15–60 μm	Granular	Cytoplasm	Conspicuously granular	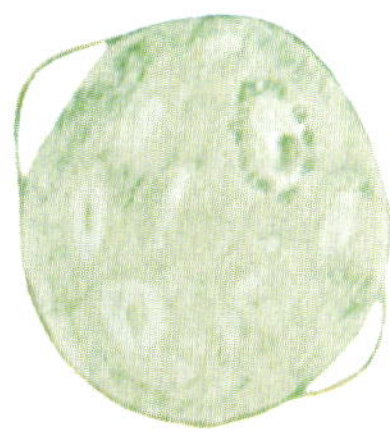 15–50 μm
	Clear finger-like	Pseudopodia	Blunt	
	Active Purposeful	Movement	***Sluggish*** Not purposeful	
	Generally invisible	Nucleus	Ring refractile granules with eccentric karyosome	
	Red blood cells (RBCs)	Inclusions	Vacuoles, crystals, vegetable cells, bacteria, ***no RBCs***	

Precyst and unripe cyst

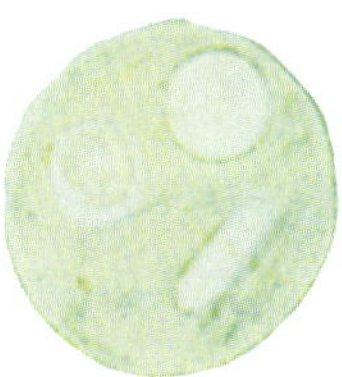	Granular	Cytoplasm	Granular	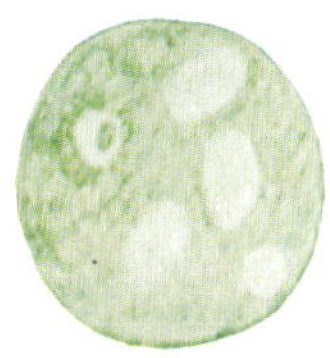
	May be refractile ring	Nucleus	Visible as refractile ring	
	Rod-like refractile chromidial bars	Inclusions	May be slender refractile chromidial bars	
	Glycogen masses		Glycogen masses	

Ripe cyst

 10–20 μm	Round	Shape	Round	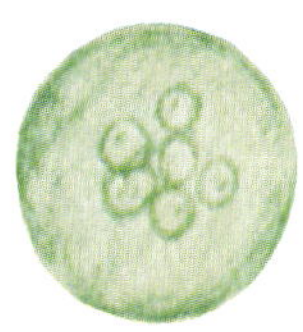 10–33 μm
	Refractile	Wall	Conspicuous refractile double outline	
	1–4 refractile nuclei with central karyosome	Nuclei	1–8 refractile nuclei with eccentric karyosomes	
	Refractile chromidial bars often present	Inclusions	Only rudimentary slender chromidial bars	

Morphology (continued)

Iodine preparations

Precyst

E. histolytica

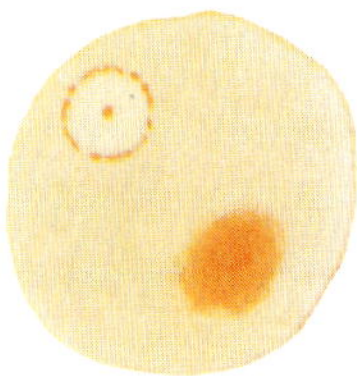

E. coli

E. histolytica		*E. coli*
Brown, diffuse	Glycogen	Brown, compact
Finely granular yellow green	Cytoplasm	Conspicuous granularity
Yellow ring with central yellow dot (karyosome)	Nucleus	Nuclear membrane with eccentric karyosome easily recognised

Stained by iron haematoxylin

Trophozoite

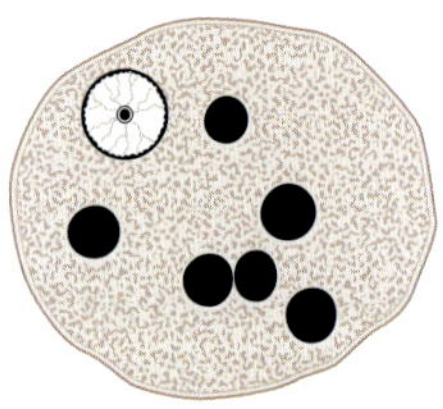

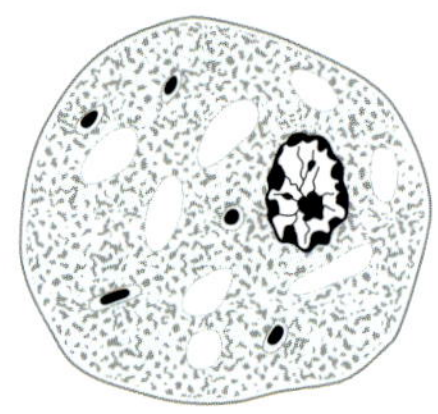

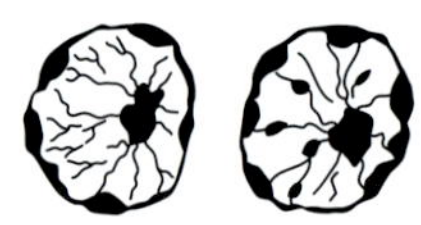

E. histolytica		*E. coli*
Purplish brown Faintly granular	Cytoplasm	Greyish blue Coarsely granular
RBC black	Inclusions	Vacuoles black, as are bacteria etc.
Lined with minute black granules	Nucleus: Membrane	Thick with plaques of black chromatin
Small black central dot	Karyosome	Eccentric black dot or plaque
Trace only seen	Fibril network	More conspicuous; may have chromatin plaques

Precyst

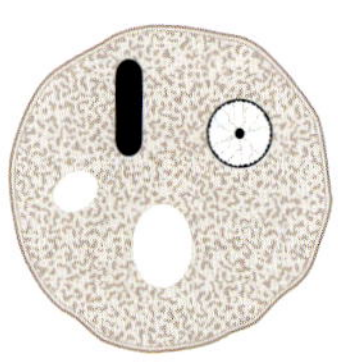

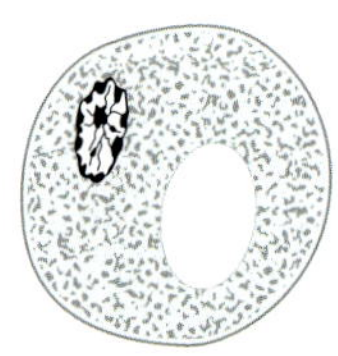

E. histolytica		*E. coli*
Round	Shape	Round
As trophozoite	Cytoplasm Nucleus	As trophozoite
Black chromidial bodies or bars	Inclusions	May have slender black chromidial bars
Glycogen (dissolved) replaced by vacuoles		Glycogen (dissolved) replaced by vacuoles

Cyst

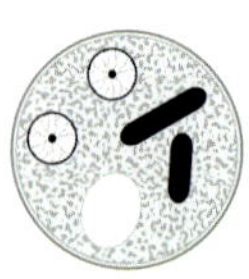

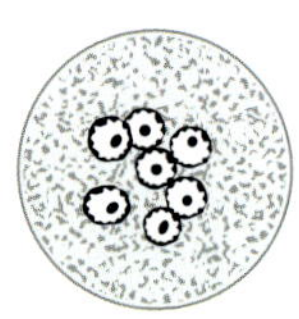

E. histolytica		*E. coli*
Grey-blue	Cytoplasm	Greyish-blue, granular
As precyst, less conspicuous or absent	Inclusions	As precyst, less conspicuous or absent In 2 nuclei stage glycogen vacuoles may be dumb-bell-shaped
Unstained, hyaline	Wall	Unstained, hyaline
As trophozoite 1–4	Nuclei	As trophozoite 1–8

Invasion of the large intestine

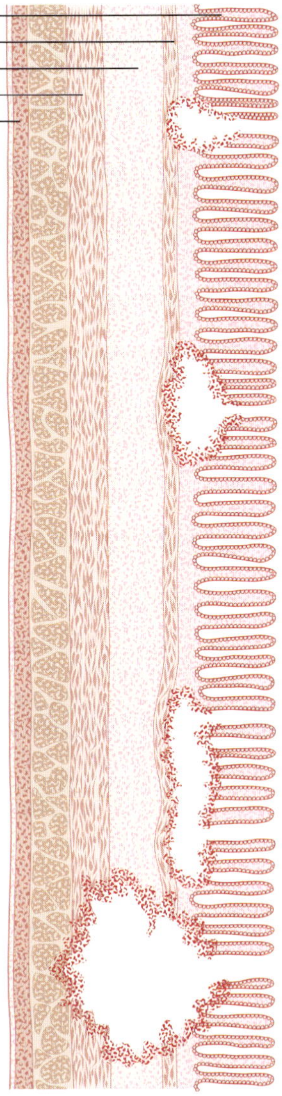

Site of entry
Initially minute then irregular ulcer shape, typically flask-like edges overhanging floor; necrotic lysed tissue; amoebae invading around discharge; necrotic debris, mucus and amoebae

Invasion elsewhere in large bowel

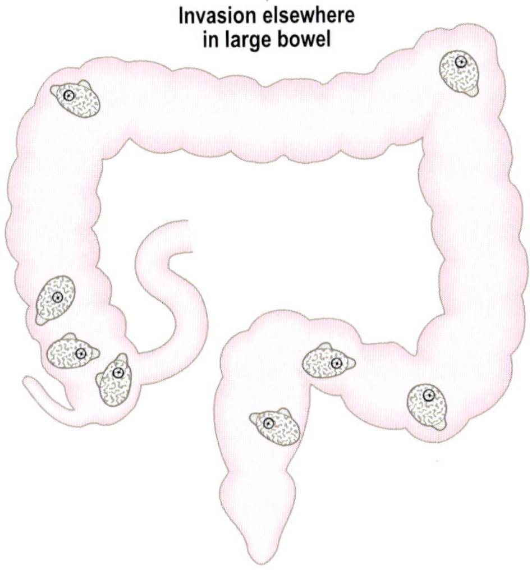

The primary ulcer	Invasion of mucosa via crypts Repair may: – overtake necrosis with healing – keep pace with necrosis causing persistent superficial lesions
'Flask-shaped'	Lag behind – extension
Extension in mucosa	Muscularis mucosae relatively resistant Accumulation of amoebae superficial to it Lateral extension of lytic necrosis
Formation of sinuses	Abscesses may coalesce under intact mucosa Later mucosa may slough with widespread ulceration
Deep extension	Muscularis mucosae eventually pierced (directly or via vessels) Deep necrosis of sub-mucosa, even muscle and sub-serosa

Complications and sequelae

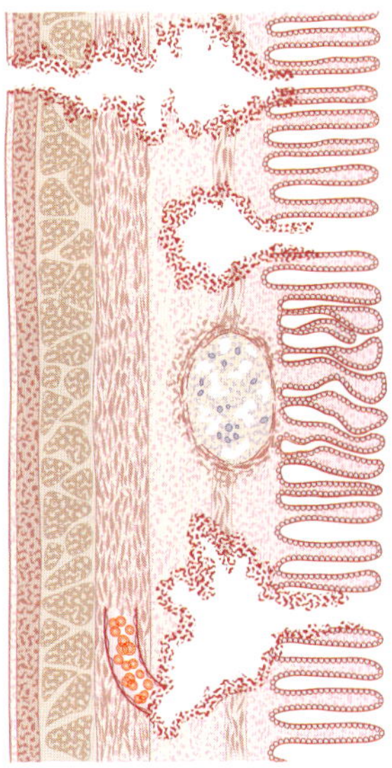

Perforation **Haemorrhage (rare)**	Peritonitis Haemorrhage
Secondary infection	Surrounding inflammatory reaction and fibroblastic proliferation
Amoeboma (rare) (Clinically simulates neoplasm) – intussusception – obstruction	A mass under oedematous mucosa with – internal abscesses of necrotic tissue and amoebae – surrounding granulomatous tissue zone with eosinophils, lymphocytes and fibroblasts – outer firm nodular fibrous tissue
Invasion of blood vessels **Direct extension outside bowel**	**Extraintestinal lesions**-page 52

Entamoeba histolytica (causing amoebiasis) (continued)

Extraintestinal lesions in amoebiasis

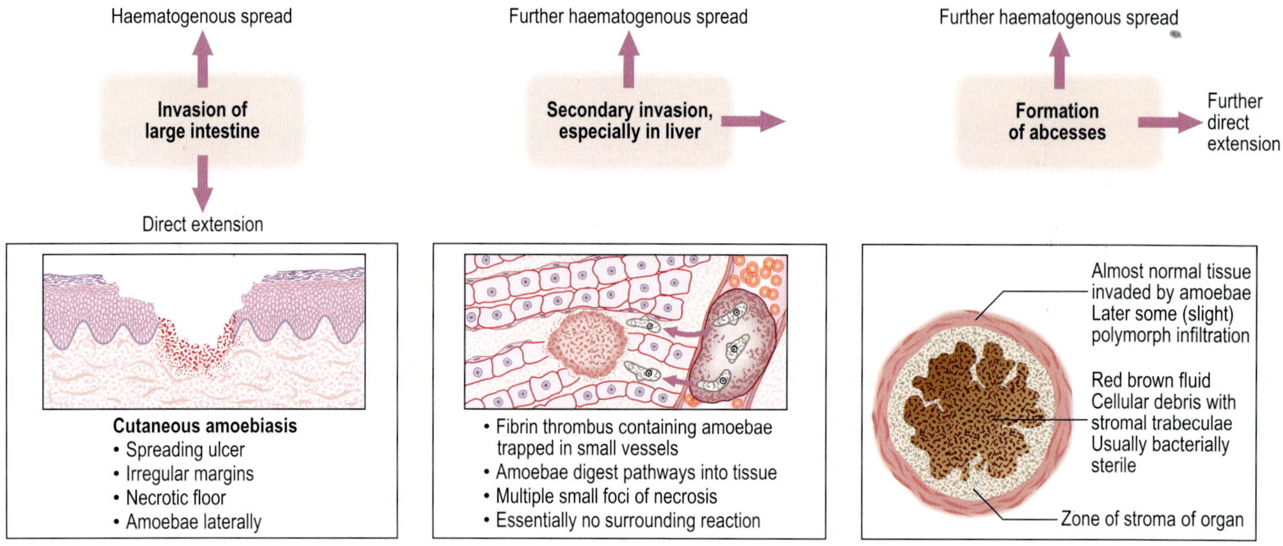

Direct extension

Haematogenous spread

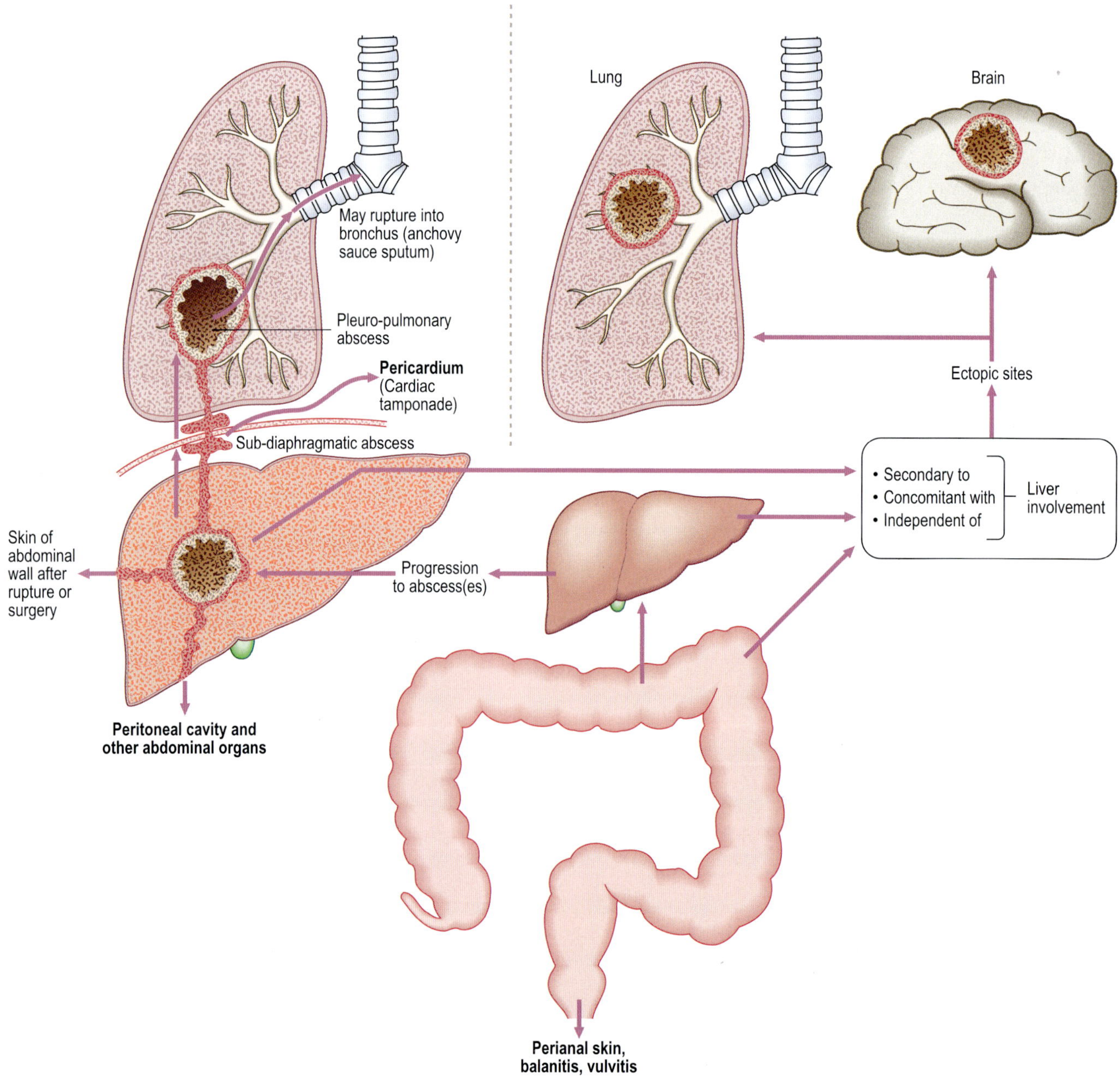

Laboratory diagnosis

Diagnosis depends primarily on demonstration of haematophagous trophozoites of *E. histolytica* in stool samples, aspirates from intestinal and other organs, biopsy material (pinch biopsy at proctoscopy or sigmoidoscopy and surgical biopsy from elsewhere) and in mucus from rectal ulcers. ELISAs are available for the detection of *Entamoeba* antigen and specific *E. histolytica* lectin antigen in faecal samples. Serology is the method of choice for diagnosis of amoebic liver disease.

Faecal appearances in amoebic dysentery

		Character
Naked eye	Faecal matter	Always present
	Mucus	Not tenacious
		Not abundant
Microscopic	1. Bacteria	Numerous
	2. Pus cells	Scanty, well preserved
	3. Red blood cells	Often in rouleaux
	4. Large macrophages	Not a feature
	5. Charcot-Leyden crystals	May be present but are non-specific
	6. Haematophagus trophozoites of *E. histolytica*	Present
Culture	Usually limited to reference or research laboratories; lectin ELISA is used for specific identification of *E. histolytica* from cyst or trophozoite. Enzyme electrophoresis from cultured trophozoite is an alternative.	

Haematophagous amoebic trophozoites.

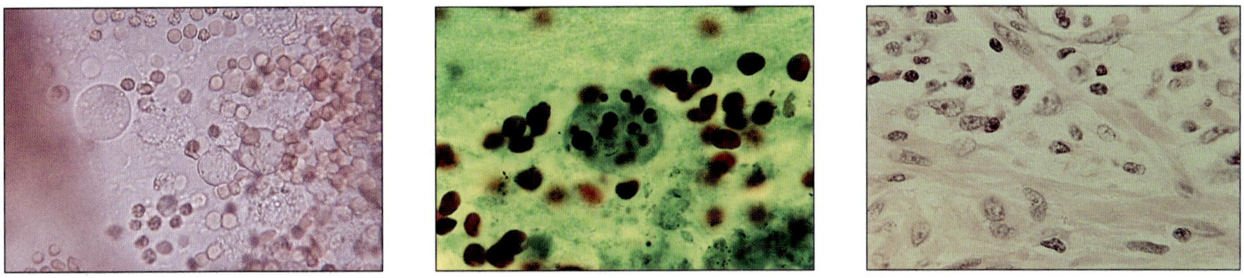

(a) Direct microscopy (b) Trichrome stain (c) H&E stain

Notes

Vegetative *E. histolytica* when seen is actively motile and moves purposefully. There are finger-like, clear pseudopodia and ingested red cells. No nucleus can be seen. Precysts or cysts found in semi-formed or solid stool have typical nuclear characteristics (1–4 nuclei) and glycogen and chromidial bars can be demonstrated.

Diagnostic tests

Polymorph leucocytosis. Examination of stool samples may show cysts and trophozoites of *E. histolytica*. Serological tests (IFAT, ELISA, cellulose acetate precipitin, latex agglutination) but serology is positive in no more than 75% of cases of amoebic colitis. Examine aspirated material for *E. histolytica*. Histology of rectal and colon biopsy material.

Other intestinal amoebae

Life cycle

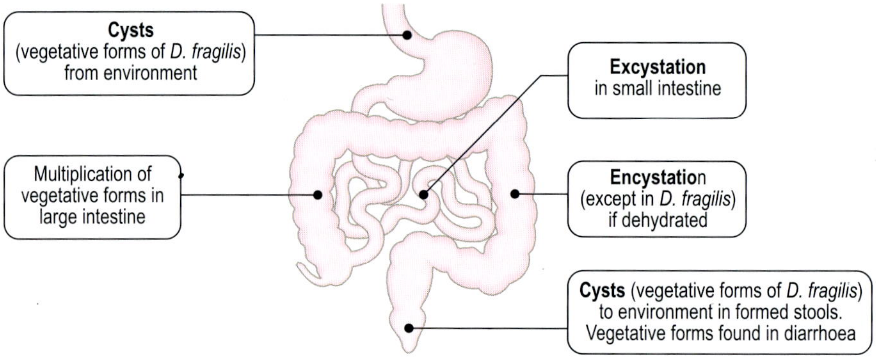

Morphology

Unstained

	Entamoeba coli	*Endolimax nana*	*Iodamoeba bütschlii*	*Dientamoeba fragilis*	*Entamoeba histolytica*	*Entamoeba dispar*	*Entamoeba hartmanni*
Vegetative forms (trophozoites)							

Size	15–50 μm	8–10 μm	8–20 μm	5–12 μm	15–60 μm	15–60 μm	15–60 μm
Motility	Sluggish	Sluggish	Fairly active	Very active	Very active	Active	Active
Ectoplasm	Little	Little	Little	Abundant	Abundant	Abundant	Abundant
Pseudopodia	Blunt, mainly granular	Blunt, mainly granular	Blunt, clear	Leaf-like, clear	Finger-like, clear	Finger-like, clear	Finger-like, clear
Endoplasm	All have granular cytoplasm with food particles, bacteria, crystals, vegetable cells, often in vacuoles. No ingested RBCs				Ingested RBCs	No ingested RBCs	No ingested RBCs
Nucleus	Ring of refractive dots	Generally invisible	Generally invisible	Two, collection of dots	Generally invisible	Generally invisible	Generally invisible
Precyst (round up, discharge food particles, bacteria, etc.)							
Glycogen	Often prominent vacuole	Rare	Conspicuous	None	Diffuse, soon disappears	Diffuse, soon disappears	Diffuse, soon disappears
Chromidial bars	Rarely seen	Rare	None	None	Large refractile bars	Large refractile bars	Large refractile bars
Cysts							

Size	10–33 μm	5–14 μm	5–18 μm	None	10–20 μm	10–20 μm	8–10 μm
Shape	Spherical or oval	Oval	Irregular	None	Spherical	Spherical	Spherical
Wall	Thick	Thin	Thin	None	Thin	Thin	Thin
Glycogen	Diffuse central	None	Well-defined vacuoles	None	Sometimes persists	Sometimes persists	Sometimes persists
Chromidial bars	Not usual	None	None	None	Sometimes present	Sometimes present	Sometimes present
Nuclei numbers	1–8	4 (at one end)	1 only	None	1–4	1–4	1–4

Stained

	Entamoeba coli	*Endolimax nana*	*Iodamoeba bütschlii*	*Dientamoeba fragilis*	*Entamoeba histolytica*	*Entamoeba dispar*	*Entamoeba hartmannii*
Cytoplasm inclusions	Stain black except glycogen as clear area	With haematoxylin, stains bluish-grey			RBCs also stain black		
Nuclear characteristics							
Membrane	Thick	Thin	Thick	Very delicate		Delicate	
Chromatin on membrane	Coarse	None	Sometimes granular	None		Fine granules	
Karyosome	Coarse, generally eccentric	Large irregular	Large lateral	Central granules		Small central	
Fibril network	May be chromatin particles	No chromatin	No chromatin	Delicate fibrils		Not often seen	
Pathogenicity	Harmless commensal	Harmless commensal	Harmless commensal	Disputed	Invasive	Harmless commensal Non-invasive	Harmless commensal Non-invasive

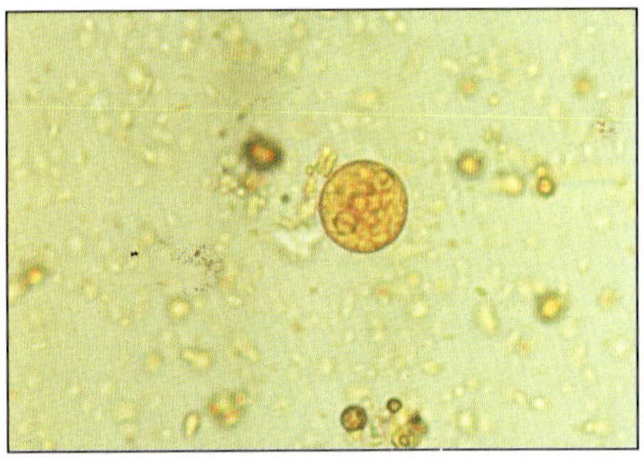

Entamoeba coli cysts

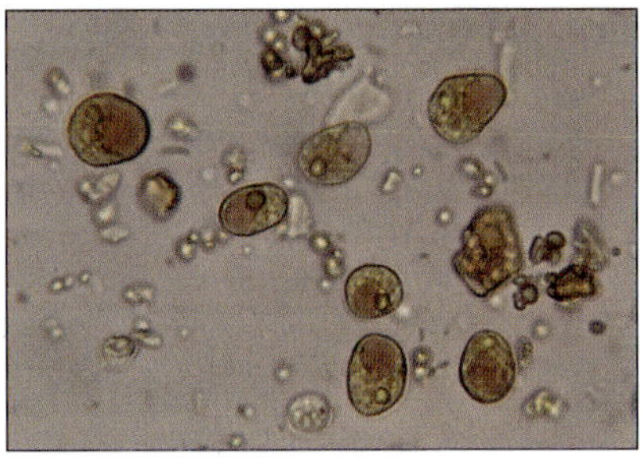

Iodamoeba bütschlii cysts

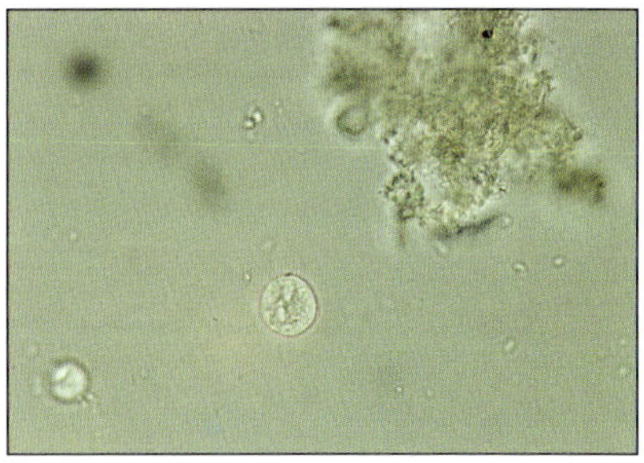

Entamoeba histolytica/dispar cysts

Intestinal flagellates

Diagnosis

Trophozoites or cysts are found in stool samples or duodenal aspirates. Duodenal string test and stool antigen detection ELISA are also possible for the detection of *Giardia*.

Distribution

These protozoa have a worldwide distribution.

Giardia intestinalis (G. lamblia)

Life cycle

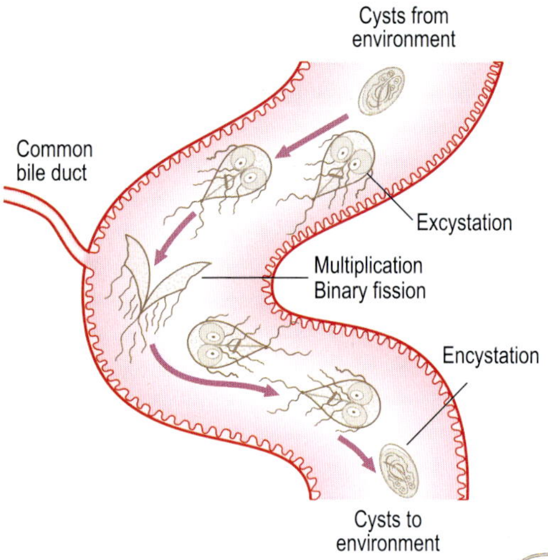

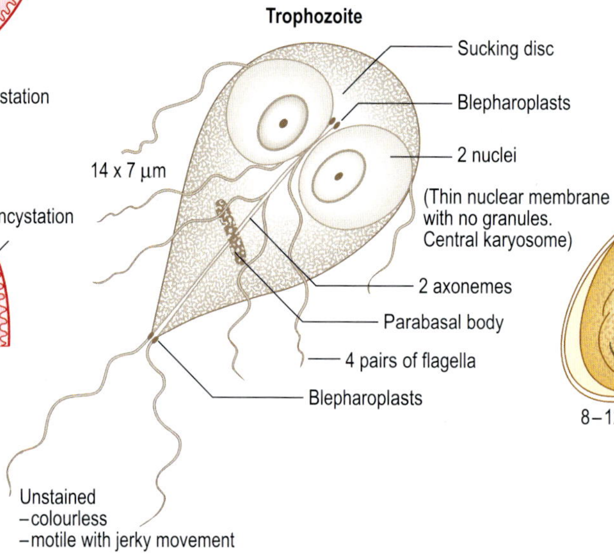

Pathogenicity
Common inhabitant of upper part of small intestine
Enteropathy, diarrhoea, steatorrhoea

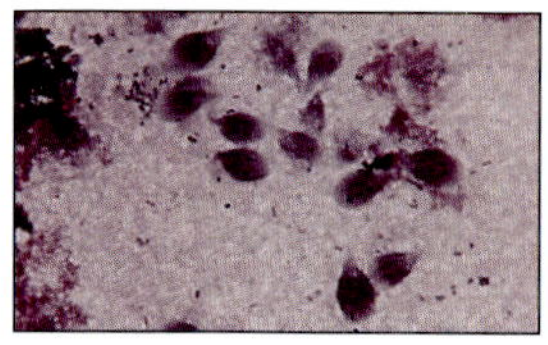

G. lamblia trophozoites

Cyst (iodine stained)

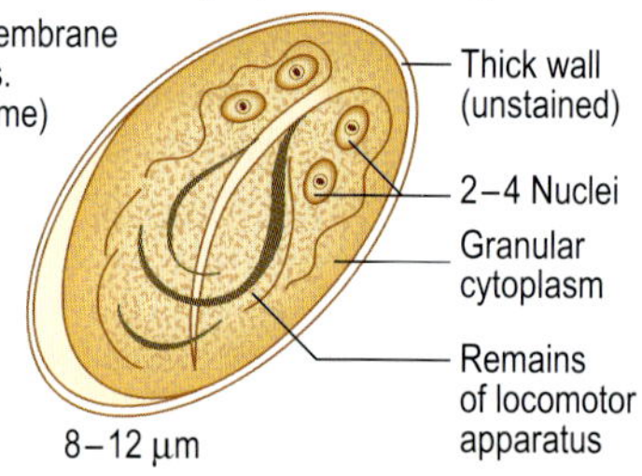

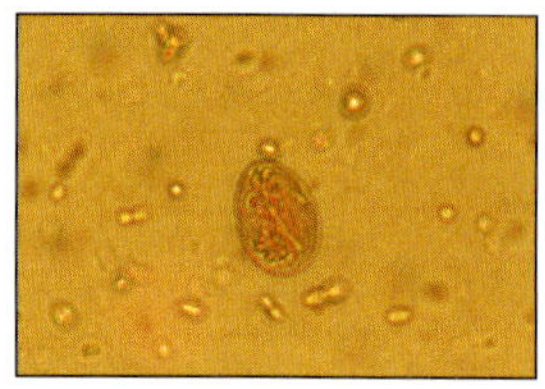

G. lamblia cyst

Chilomastix mesnili

Life cycle

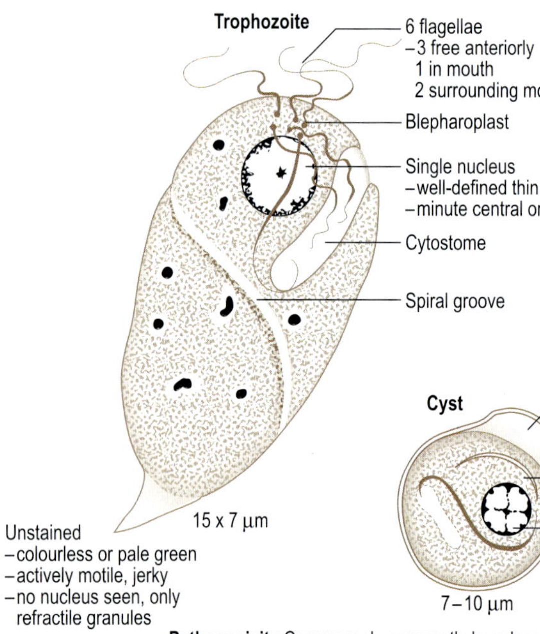

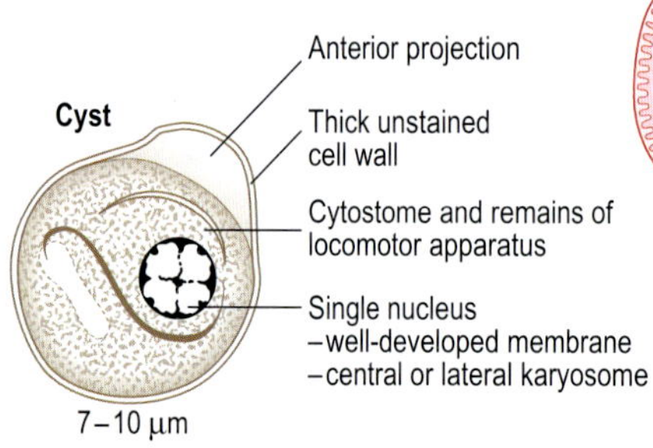

Pathogenicity Commensal–apparently harmless

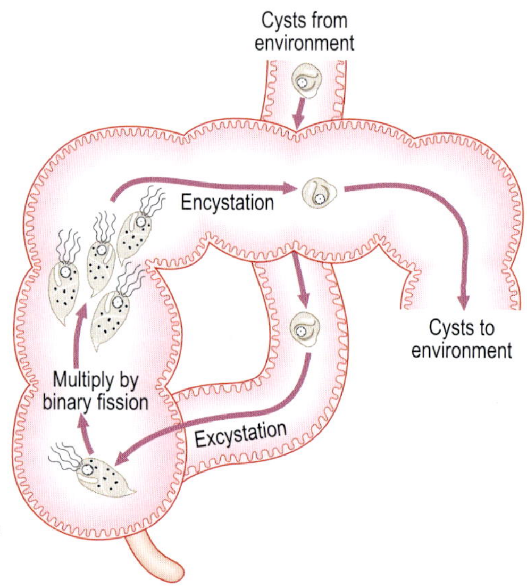

Trichomonas species

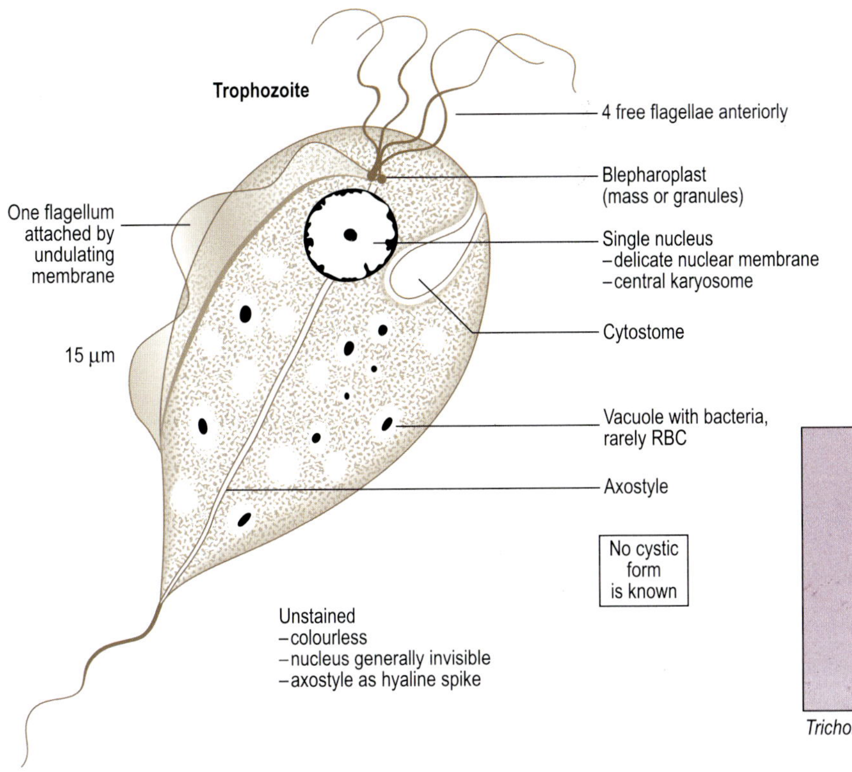

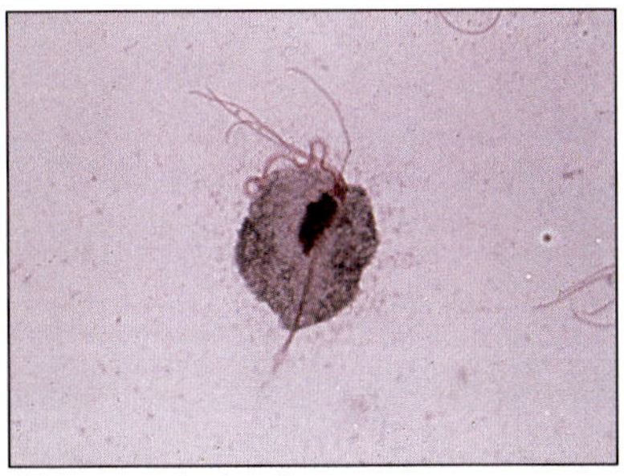

Trichomonas trophozoite

T. hominis

This is illustrated above. The trophozoite inhabits the small and large intestine. There is no proof as yet that it has any pathogenicity.

T. vaginalis

Morphologically this is the same as *T. hominis* (above) but there is no free posterior flagellum beyond the undulating membrane. There is a marked parabasal body. It inhabits the urethra in the male and the vagina in the female, and is a cause of urethritis and vaginitis.

Demonstration of *T. vaginalis* is made by direct microscopy or after staining with acridine orange fluorescence stain. Cultures can be made using Feinberg–Whittington or Diamond's medium.

Intestinal ciliates

Balantidium coli

Found in South and Central America, parts of Asia and some Pacific islands.

In its vegetative state, recognizable by the oval shape, coarse cilia, contractile vacuoles and the horseshoe- or kidney-shaped macronucleus. Reproduction is by binary fission.

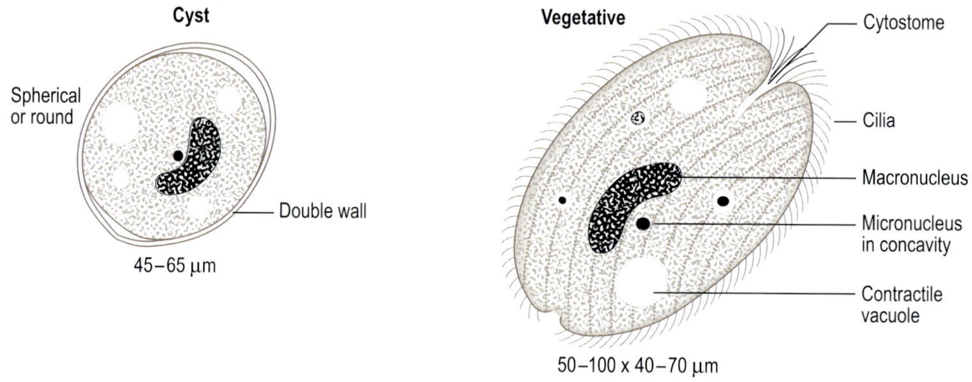

Life cycle

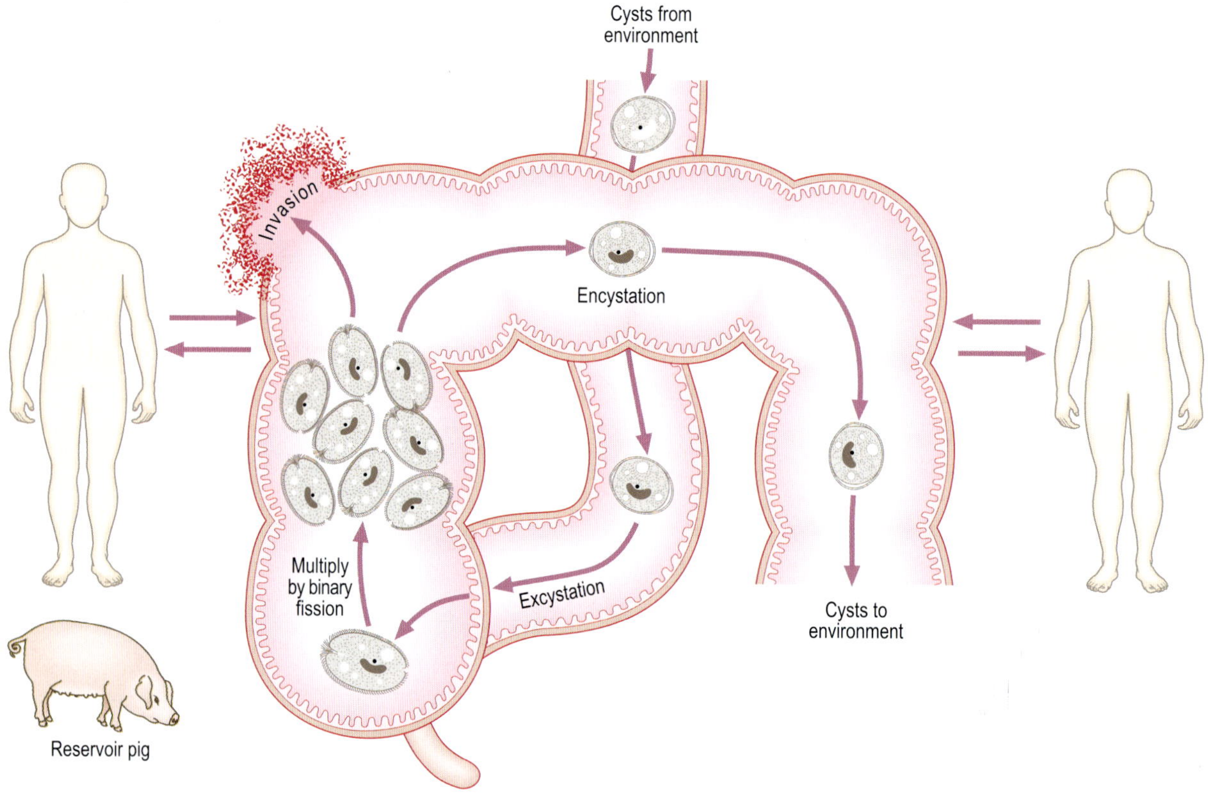

Pathology and Clinical features

Problems occur in the ileum, colon and rectum but there is no extraintestinal spread. The parasite is a cause of dysentery, although the ulcers are wider mouthed than those of amoebic dysentery. Secondary infection is frequent. The main complication is perforation.

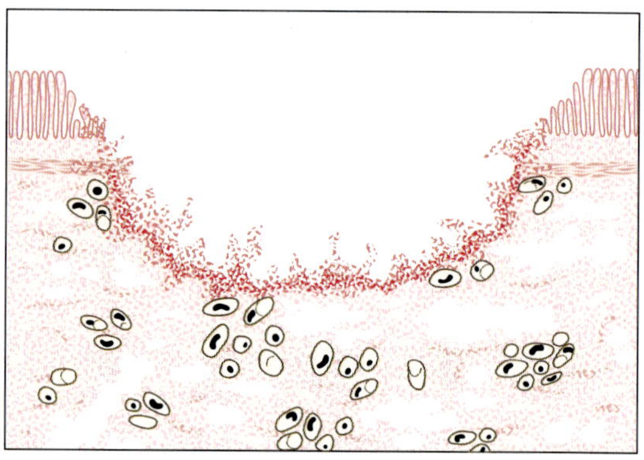

Laboratory diagnosis

Trophozoites are found in diarrhoea and, in a fresh specimen, can be seen in active rotational movement. Cysts are found in semi-formed and formed stools.

Tissue protozoa

Toxoplasma gondii

Toxoplasma has a very wide mammalian host range.

Morphology

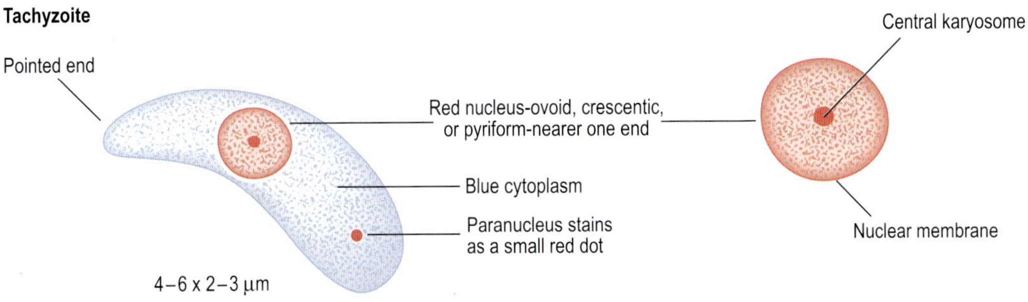

Habitat

Tachyzoites: single (free or intracellular) or in masses (pseudocysts)

In nucleated cells, especially macrophages

Bradyzoites (similar to tachyzoites but less active metabolically) in tissue cysts

Life cycle

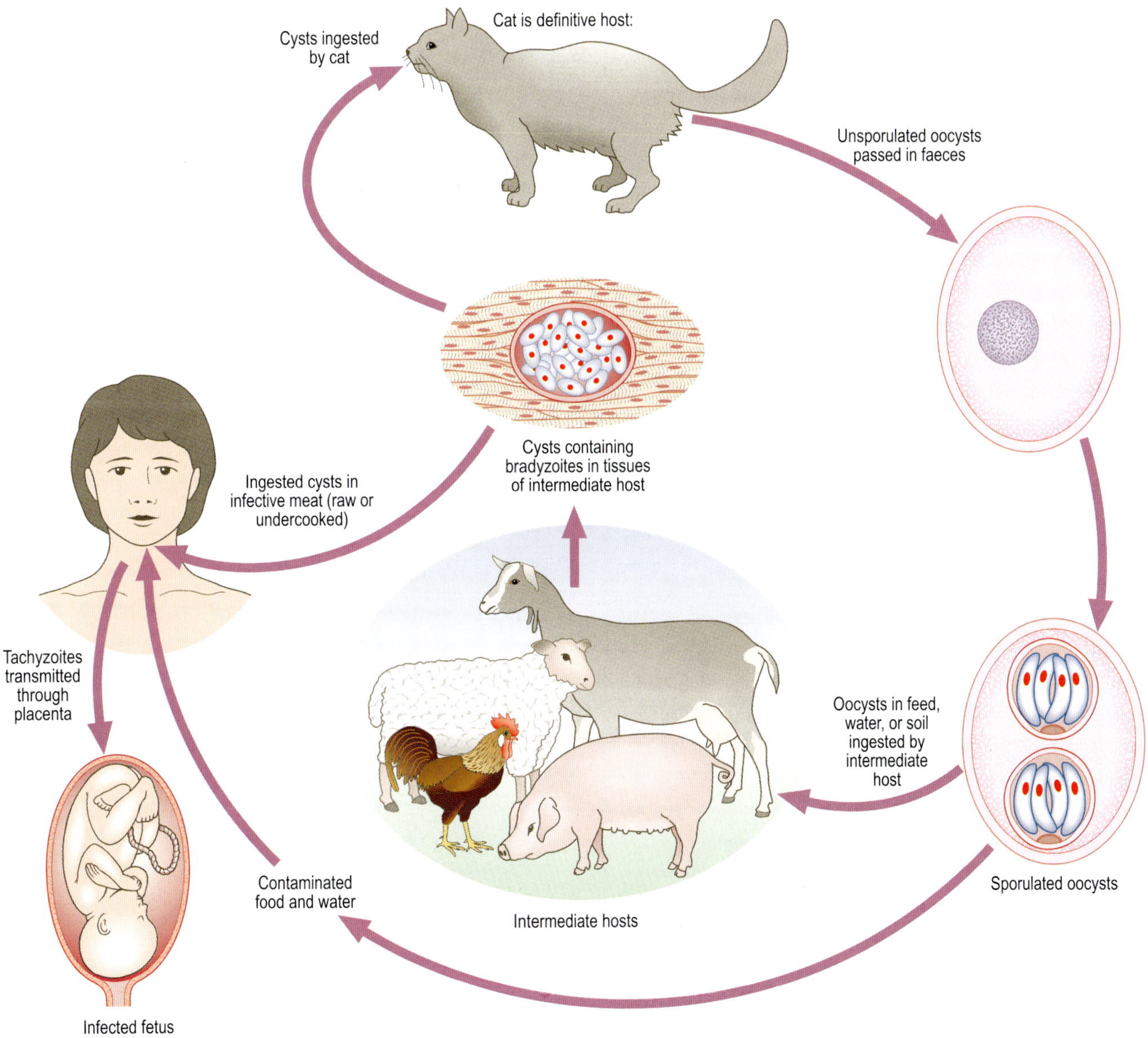

Pathology and Clinical features

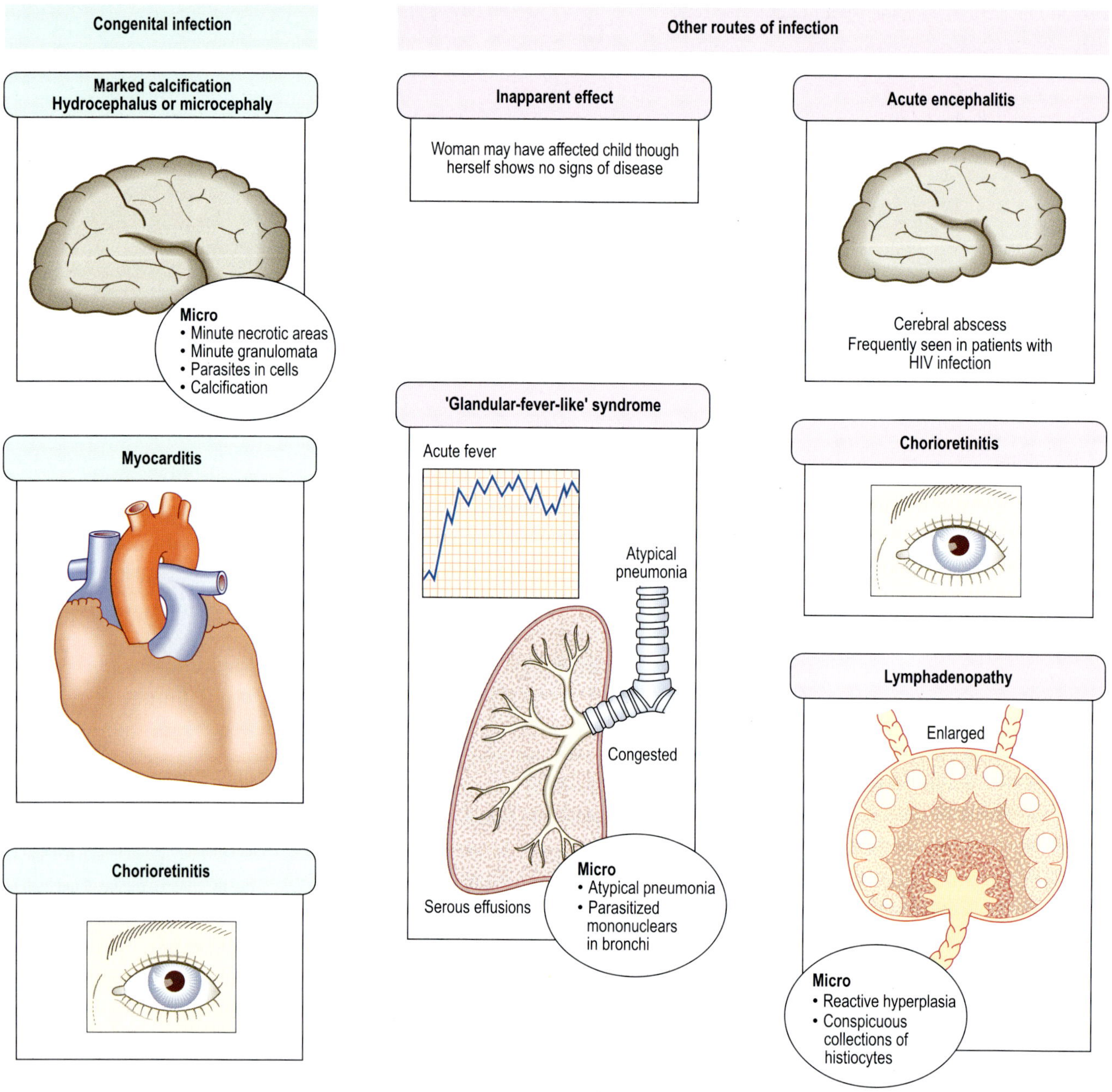

Laboratory diagnosis

Diagnosis is usually made serologically by demonstration of specific antibodies. Methods include Latex agglutination, ELISA and ISAGA. The 'gold standard' for *Toxoplasma* serological diagnosis is the Sabin-Feldman dye test.

Lymph node biopsy should not be required to diagnose *Toxoplasma* but if performed because another diagnosis was suspected, the findings are as stated above.

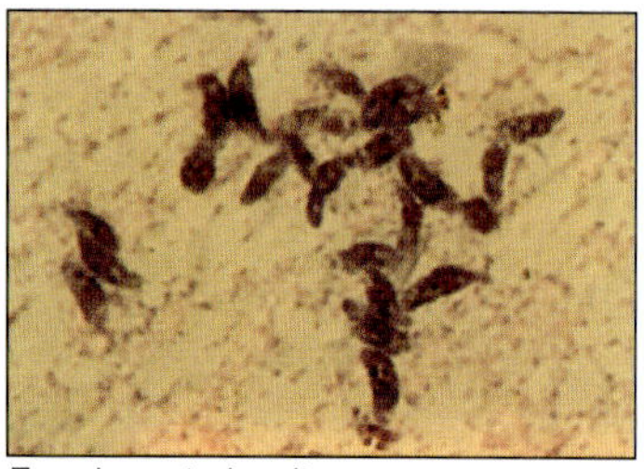

Toxoplasma tachyzoites

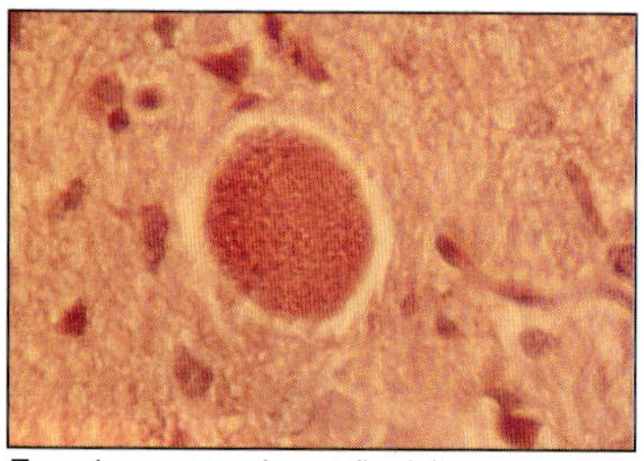

Toxoplasma pseudocyst (brain)

Malaria parasites

Classification

Class	**Haematozoea**
Order	**Haemosporida** • Sexual and asexual generations in different hosts • Parasitic in fixed tissue cells and RBCs of vertebrate host
Family	**Plasmodiidae** • Include human malaria parasites • Produce pigment in asexual cycle in RBCs of vertebrates • Produce gametocytes in RBCs of vertebrates • Sporogony (sexual cycle) in invertebrates
Genus	**Plasmodium** • Schizogony (asexual cycle) in: –RBCs –other tissue cells of vertebrate host • Pigment derived from haemoglobin of infected RBC • Gametocytes develop in some RBCs. These undergo sporogony (sexual cycle) in female anopheline mosquitoes • Sporozoites produced in mosquito, infective to vertebrate host • All malaria parasites included in this genus

Class	***P. vivax***	***P. malariae***	***P. falciparum***	***P. ovale***
	Causes benign tertian malaria	Causes quartan malaria	Causes malignant tertian malaria	Causes ovale malaria

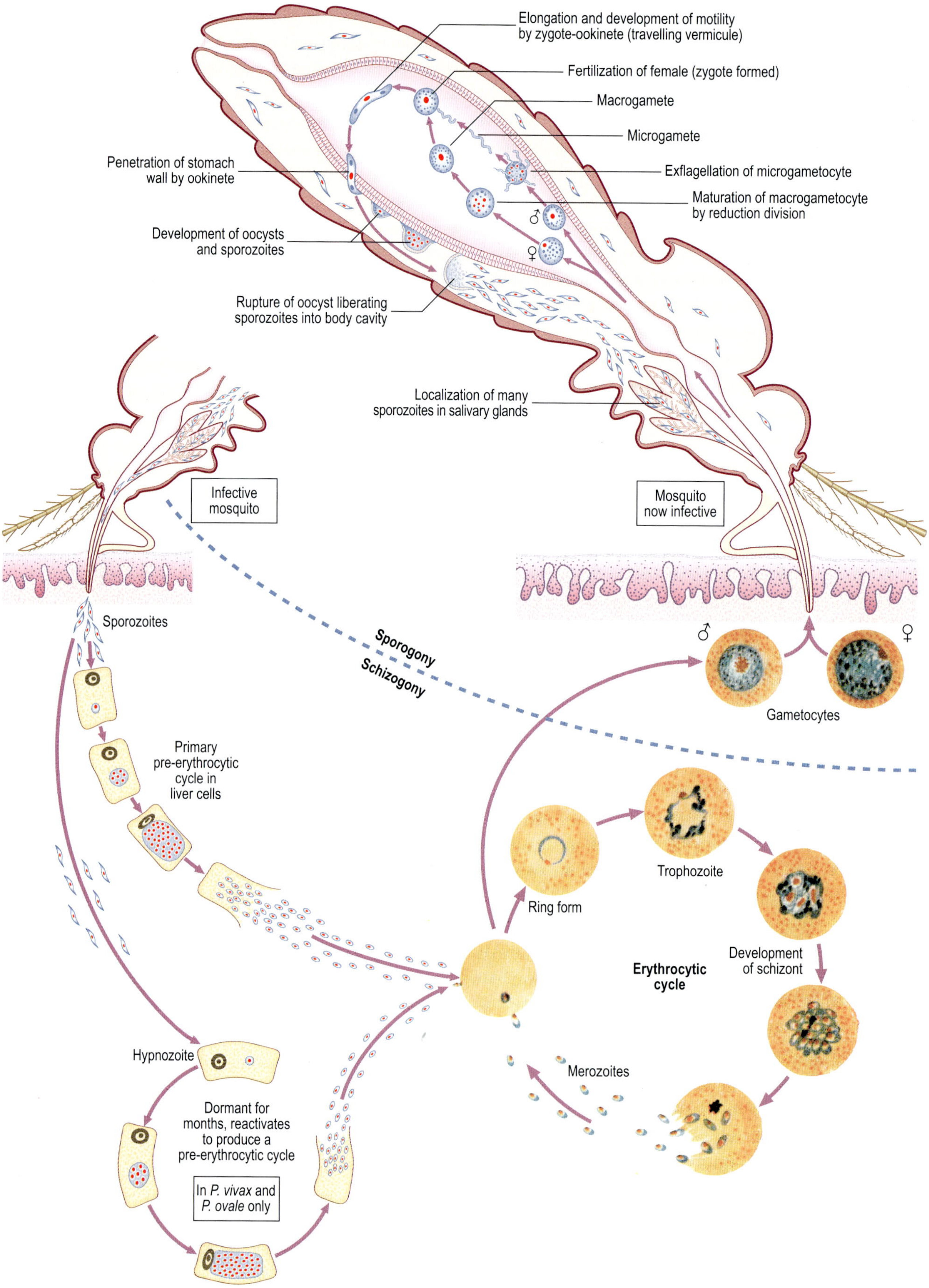
Elongation and development of motility by zygote-ookinete (travelling vermicule)
Fertilization of female (zygote formed)
Macrogamete
Microgamete
Exflagellation of microgametocyte
Maturation of macrogametocyte by reduction division
♂
♀
Penetration of stomach wall by ookinete
Development of oocysts and sporozoites
Rupture of oocyst liberating sporozoites into body cavity
Localization of many sporozoites in salivary glands
Infective mosquito
Mosquito now infective
Sporozoites
Sporogony
Schizogony
♂
♀
Gametocytes
Primary pre-erythrocytic cycle in liver cells
Trophozoite
Ring form
Erythrocytic cycle
Development of schizont
Merozoites
Hypnozoite
Dormant for months, reactivates to produce a pre-erythrocytic cycle
In P. vivax and P. ovale only

Morphology

Stained by Leishman or Giemsa

Schizogony (asexual cycle)

Pre-erythrocytic cycle in liver cells

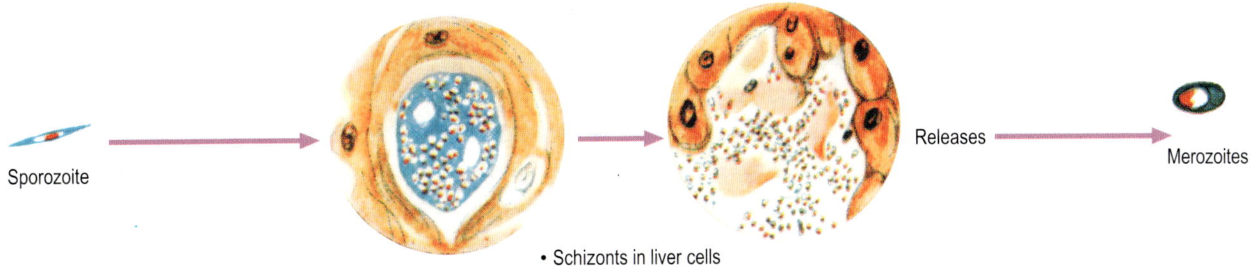

- Schizonts in liver cells
- No pigment at this stage

Erythrocytic stage in RBCs

General features

The parasite

- Cytoplasm blue
- Chromatin red
- Pigment (from haemoglobin) varies in colour and time of appearance

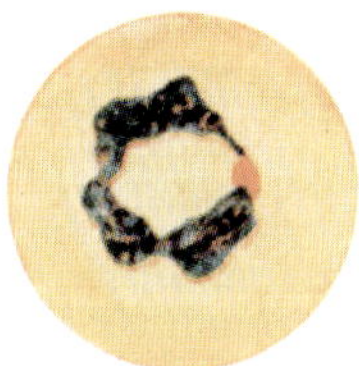

The red cell

May vary in :

- Size

Normal RBC

- Shape

May contain :

Pink spots in cytoplasm unoccupied by parasite

- Schüffner's or James' dots

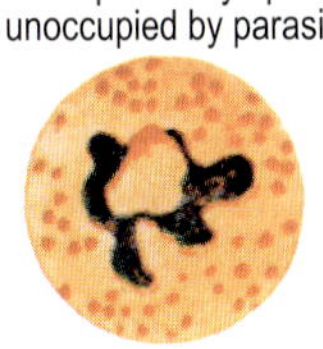

- Maurer's clefts

Brick red clefts in cytoplasm

RBC characteristics	*P. vivax*	*P. malariae*	*P. falciparum*	*P. ovale*
Size	Larger than mature RBC	Smaller, older RBC	Mature RBC	Larger than mature RBC
Colour	Pale	Normal	Normal	Pale
Shape	Round	Round	Round May be crenated	Oval May be fimbriated
Cytoplasmic inclusions	Schüffner's dots present	None	Maurer's clefts may be present in late trophozoites	James' dots conspicuous

Stages in thin films

Ring forms (early trophozoites)

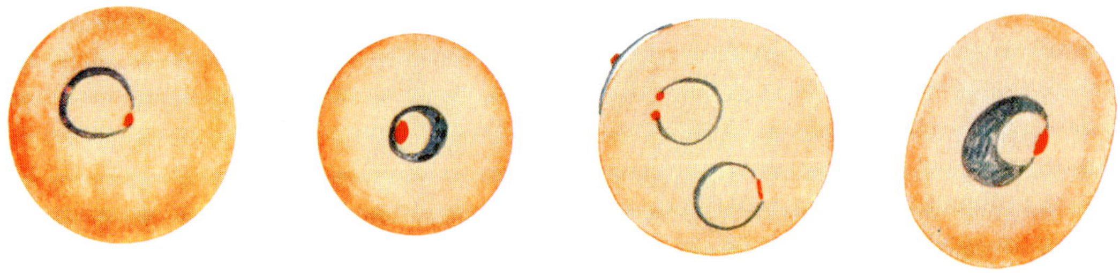

	P. vivax	*P. malariae*	*P. falciparum*	*P. ovale*
Size	1/3 RBC	Up to 1/3 RBC	1/3 RBC	1/3 RBC
Shape	Delicate ring	Compact ring	Very delicate ring	Dense ring

	P. vivax	*P. malariae*	*P. falciparum*	*P. ovale*
Chromatin	Fine dot	One mass often inside ring	Fine dots Frequently two	Dense, well-defined mass
Accolé forms*	Sometimes	None	Frequent	None
Pigment	None at this stage	May be present	None at this stage	None at this stage
Multiple parasitized cells	Sometimes	Rare	Frequently with high parasitaemia	Rare

* Forms situated on margin of RBC

Developing trophozoites

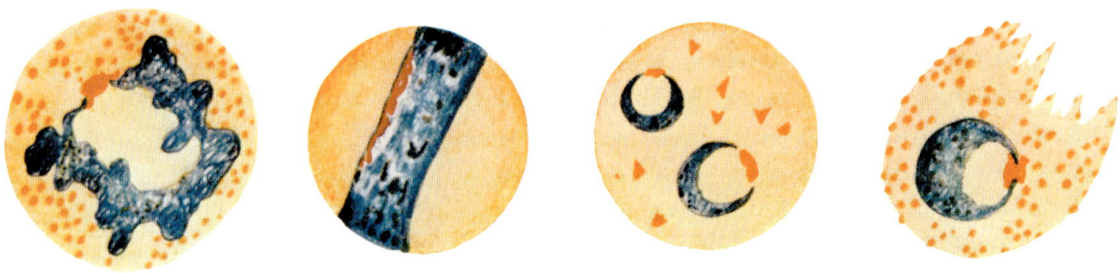

	P. vivax	*P. malariae*	*P. falciparum*	*P. ovale*
Size	Large	Small, but appears large relative to size of RBC	Small	Small
Shape	Very irregular, amoeboid	Compact, often band forms	Compact, with cytoplasmic vacuolation	Compact
Chromatin	Dots or threads	Prominent, often as a band	Dots or threads	Large irregular clumps
Pigment				
—texture	Fine	Coarse	Coarse	Coarse
—colour	Yellow brown	Dark brown	Black	Dark yellow brown
—quantity	Medium	Abundant	Medium	Medium
—distribution	Scattered fine particles	Scattered clumps and rods	Aggregated in one or two clumps	Scattered coarse particles

Immature schizonts

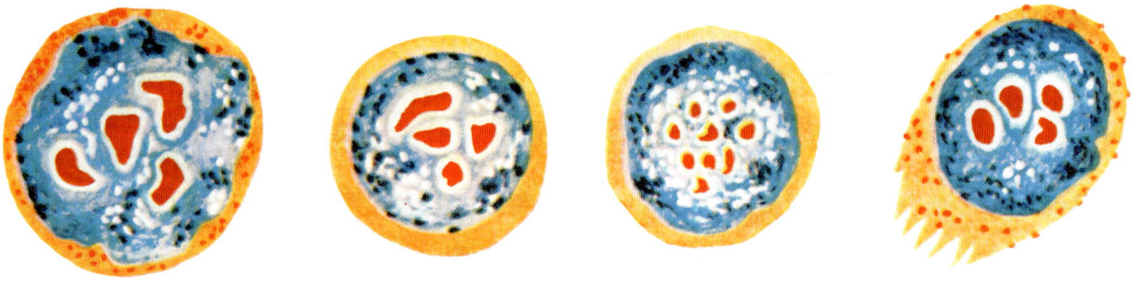

(rarely seen in peripheral blood)

	P. vivax	*P. malariae*	*P. falciparum*	*P. ovale*
Size	Almost fills RBC	Almost fills RBC	Almost fills RBC	Almost fills RBC
Shape	Somewhat amoeboid	Compact	Compact	Compact
Chromatin	Numerous irregular masses	Few irregular masses	Irregular masses	Few irregular masses
Pigment	Scattered	Scattered	Single clump	Scattered

Mature schizonts

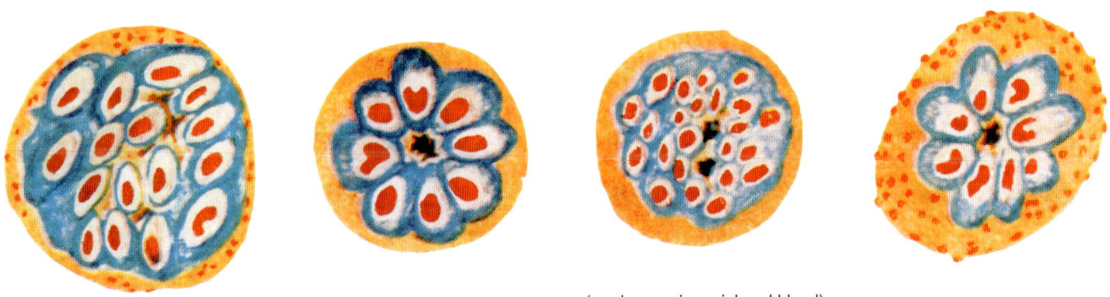

(rarely seen in peripheral blood)

	P. vivax	*P. malariae*	*P. falciparum*	*P. ovale*
Size	Fills RBC	Nearly fills RBC	Nearly fills RBC	Fills 3/4 RBC
Shape	Segmented	Segmented daisy head	Segmented	Segmented
Merozoites				
—range	14–24	6–12	8–32	6–12
—mean	16	8	24	8
—size	Medium	Large	Small	Large
Pigment	Aggregated in centre (yellow brown)	Aggregated in centre (dark brown)	Aggregated in centre (black)	Aggregated in centre (dark yellow brown)

Stages in thin films (continued)

Microgametocytes (male)

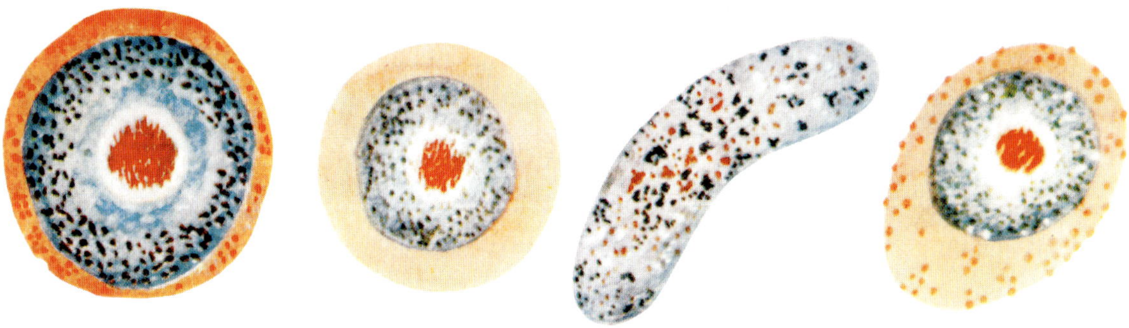

	P. vivax	*P. malariae*	*P. falciparum*	*P. ovale*
Time of appearance	3–5 days	7–14 days	7–12 days	12–14 days
Number in bloodstream	Many	Scanty	Many	Scanty
Size	3/4 fills RBC	1/2 to 2/3 fills RBC	Larger than RBC	1/2 to 2/3 fills RBC
Shape	Round or oval compact	Round compact	Kidney-shaped Bluntly round ends	Round compact
Cytoplasm	Pale blue	Pale blue	Reddish blue	Pale blue
Chromatin	Single chromatin mass	*As for P. vivax*	Fine granules scattered throughout	*As for P. vivax*
Pigment	Abundant brown granules throughout	*As for P. vivax*	Dark granules throughout	*As for P. vivax*

Macrogametocytes (female)

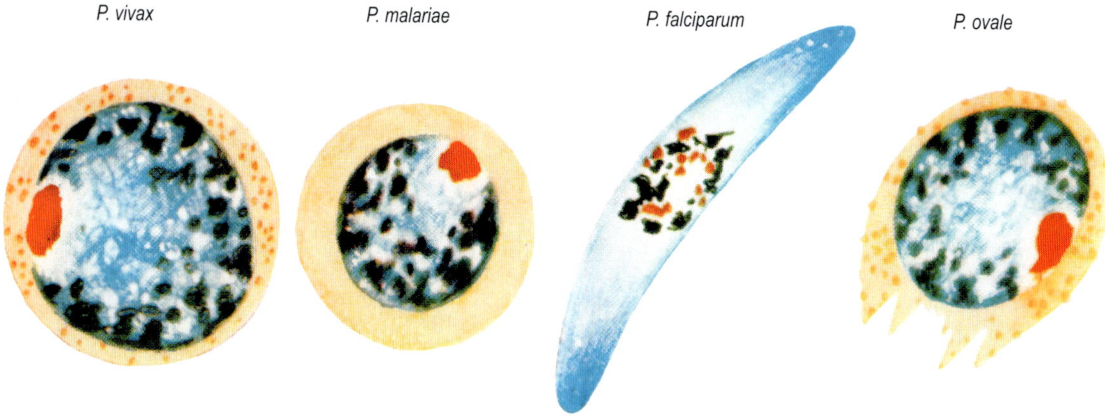

	P. vivax	*P. malariae*	*P. falciparum*	*P. ovale*
Time of appearance	3–5 days	7–14 days	7–12 days	12–14 days
Number in bloodstream	Many	Scanty	Many	Scanty
Size	3/4 fills RBC	1/2 to 2/3 fills RBC	Larger than RBC	1/2 to 2/3 fills RBC
Shape	Round or oval compact	Round compact	Crescentic-sharply rounded or pointed ends	Round compact
Cytoplasm	Dark blue	Dark blue	Dark blue	Dark blue
Chromatin	Compact peripheral mass	*As for P. vivax*	Compact masses near centre	*As for P. vivax*
Pigment	Small masses round periphery	*As for P. vivax*	Black, rod-like granules round nucleus	*As for P. vivax*

Morphology in stained thick films

Note that the parasites are not flattened in the film and so appear smaller than in thin film. The red cells are haemolyzed in processing so there is no guide to the size, shape or colour of the RBCs. Schüffner's dots are indefinite and there are no Maurer's clefts.

P. vivax

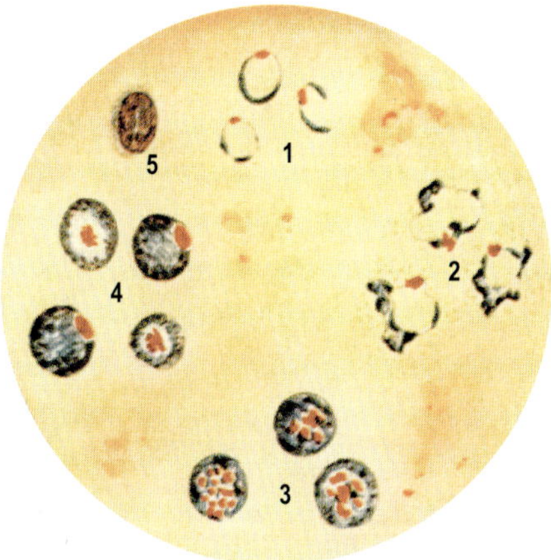

1. Ring forms, small fine rings often broken
2. Trophozoites, markedly irregular cytoplasm
3. Schizonts, many (average 16) small merozoites
4. Gametocytes, compact parasites with features of ♂ and ♀ as described
5. White blood cell

P. malariae and P. ovale

Almost identical but James' dots may be visible in the latter

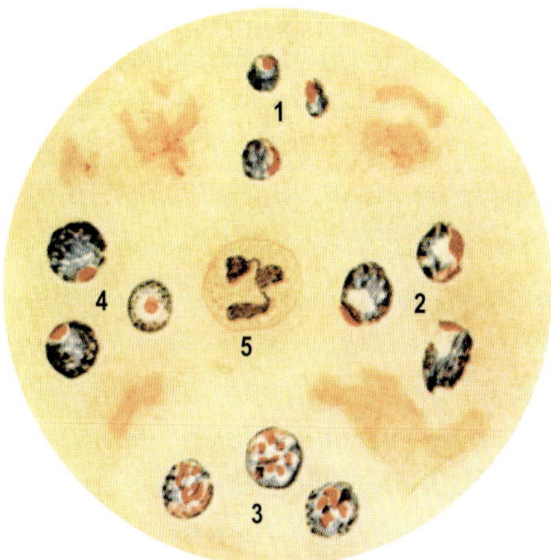

1. Ring forms, compact rings
2. Trophozoites, solid regular cytoplasm
3. Schizonts, few (average 8) large merozoites
4. Gametocytes, very difficult to distinguish from *P. vivax*
5. White blood cell

P. falciparum

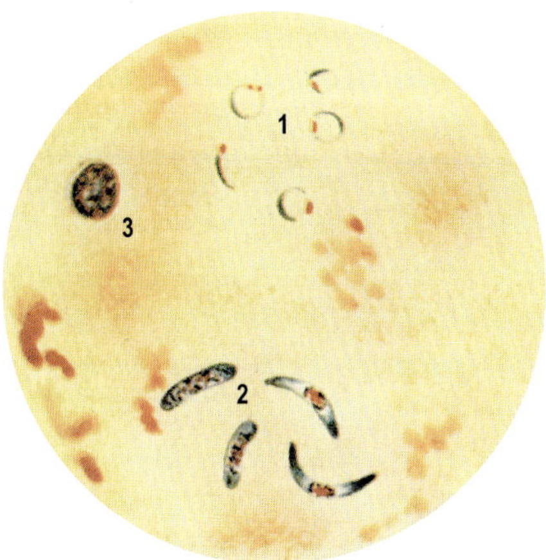

1. Ring forms, very small, fine rings usually unbroken trophozoites (with vacuolated cytoplasm) and schizonts are rarely seen in peripheral blood
2. Gametocytes, characteristic crescentic ♂ and ♀ forms
3. White blood cell

Pathology and Clinical features

Plasmodium vivax, *P. ovale*, *P. malariae* and uncomplicated *P. falciparum* malaria have similar features with fever, rigors, headache, muscle aches, malaise and anorexia. Anaemia may develop and the liver and spleen may become enlarged. Because the clinical appearances are non-specific, malaria may be misdiagnosed, e.g. as a viral infection, with severe consequences.

Plasmodium falciparum infection can readily progress to severe malaria, the clinical criteria of which have been defined by a World Health Organisation working group. One or more of the following features in the presence of asexual parasitaemia indicate severe falciparum malaria: • cerebral malaria • severe anaemia • renal failure • pulmonary oedema or adult respiratory distress syndrome • hypoglycaemia • circulatory collapse or shock • spontaneous bleeding from the gums, nose, gastrointestinal tract and/or laboratory evidence of disseminated intravascular coagulation • repeated generalised convulsions (more than two in 24 hours despite cooling) • acidaemia (arterial pH < 7.25) or acidosis (plasma bicarbonate < 15 mmol/L) • macroscopic haemoglobinuria.

Other features of severe falciparum malaria include impaired consciousness less severe than coma, prostration, hyperparasitaemia, jaundice and hyperpyrexia.

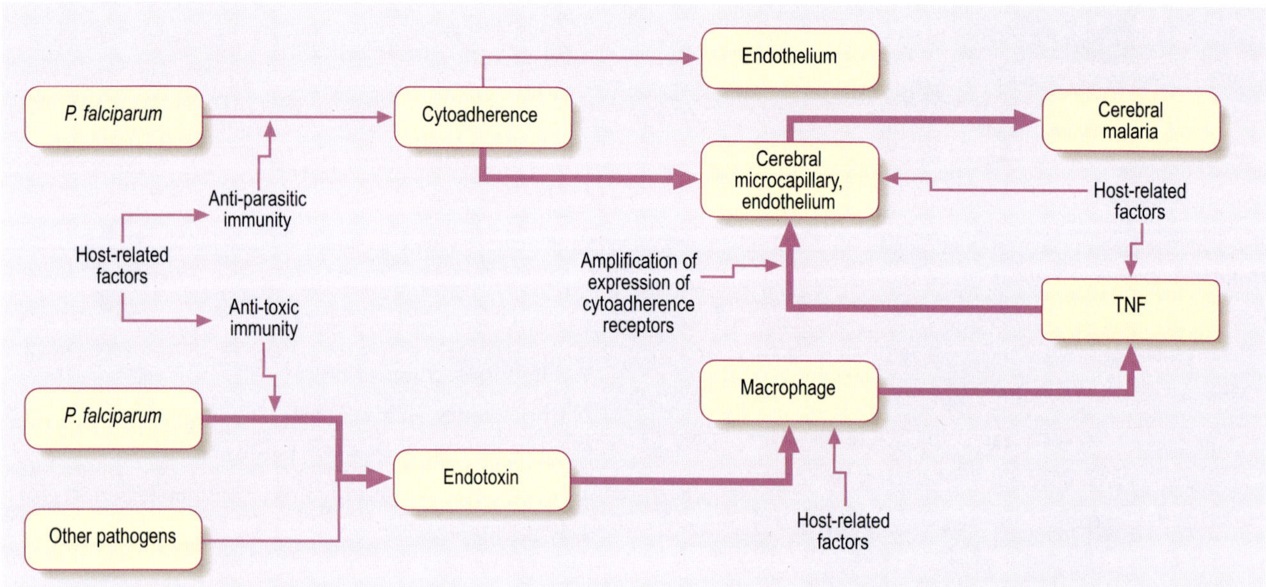

Sequence of events leading to cerebral malaria

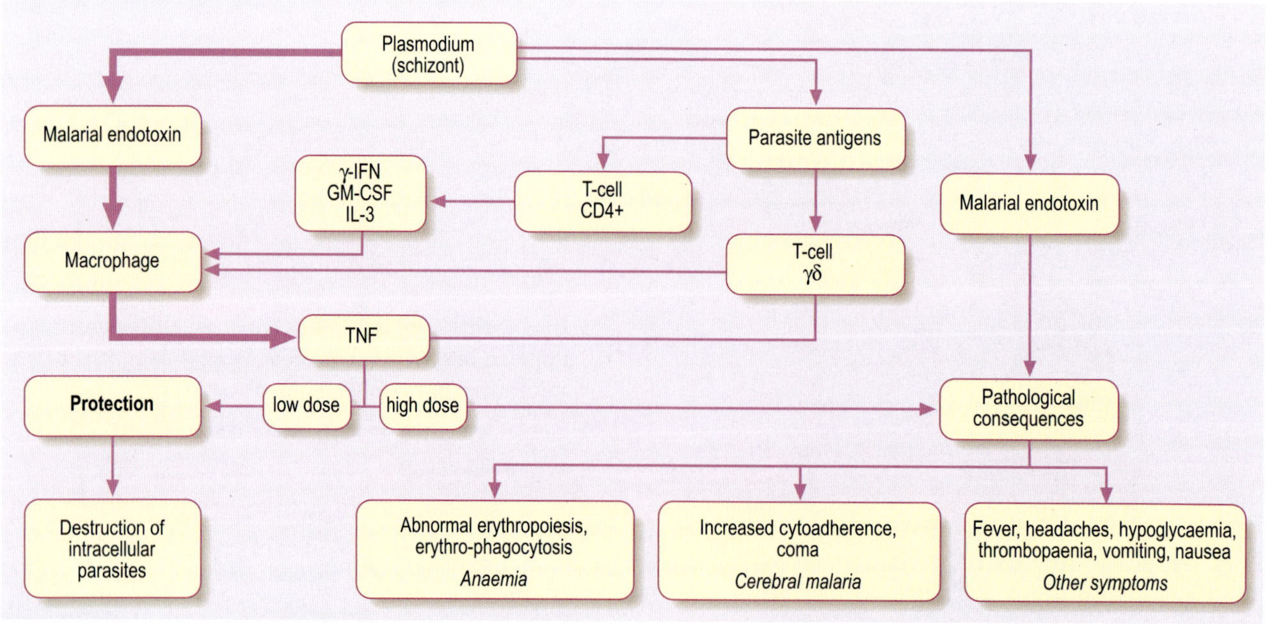

Role of TNF

(Both of the above figures after Figs 20.17 and 20.18 in Hommel M, Gilles H M, *Malaria* (Chapter 20). In Cox E G, Kreier J P, Wakelin D, eds. Topley and Wilson's *Microbiology and Microbial Infections*, Vol 5. Parasitology. London: Arnold; 1998)

Laboratory diagnosis

Malaria parasites in thin blood film. Stained by Leishman or Giemsa at pH 7.2

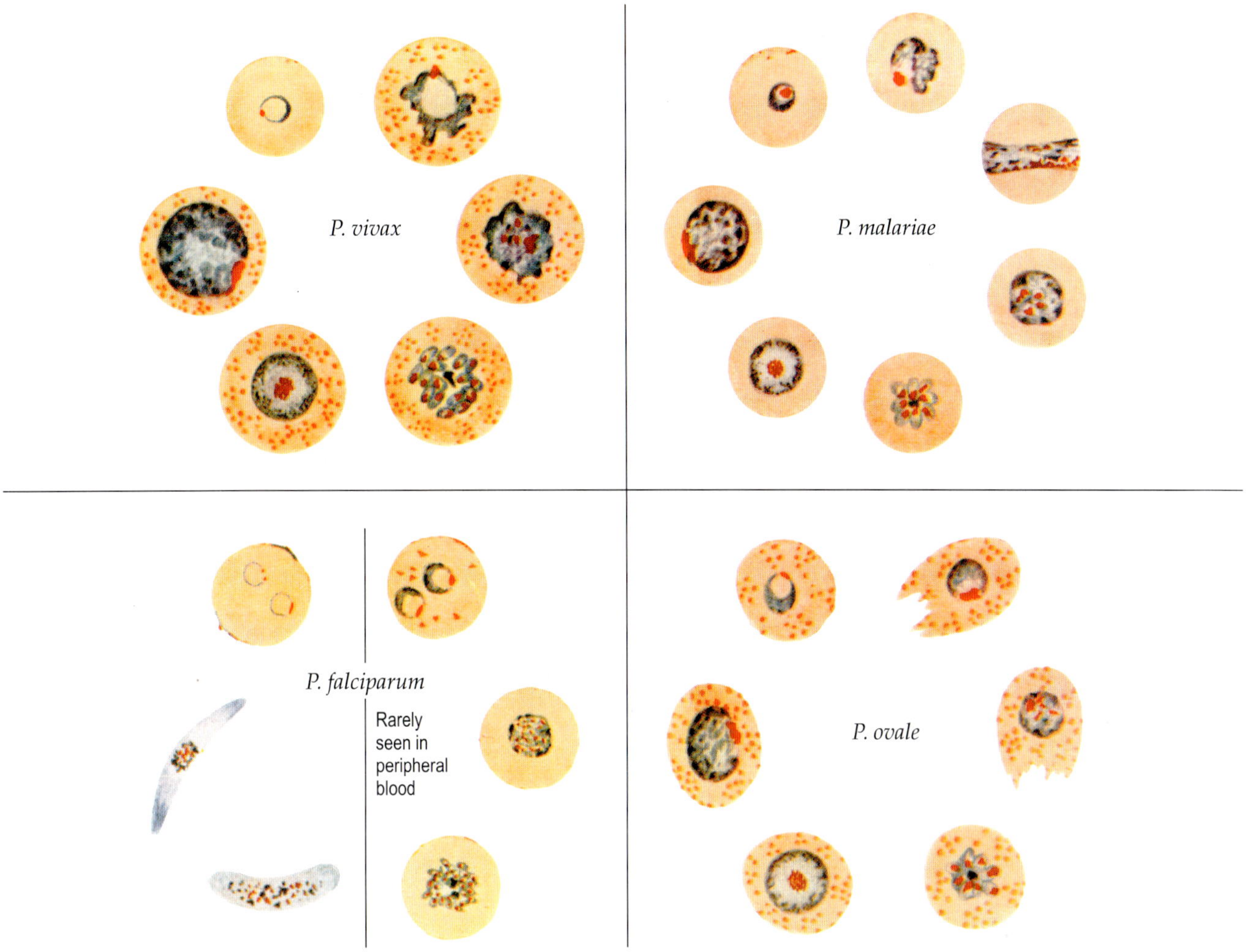

It is also possible to use thick blood films stained by Field or Giemsa. Bone marrow films may also be examined. Serology (IFAT or ELISA) is not appropriate for the detection of acute malaria but is deployed as a retrospective test for epidemiological use to establish the cause of nephrotic syndrome or hyperreactive malarial splenomegaly (HMS).

Antigen Detection

P. falciparum expresses a specific antigen HRP2 on the surface of the parasitized RBC. This can be detected by using immunochromatographic antigen capture techniques (AMRAD ICT, Becton Dickinson ParaSight F). Parasite lactate dehydrogenase (pLDH) is biochemically and antigenically distinct from human LDH and is produced by all *Plasmodium* species. Gold-labelled monoclonal and polyclonal antibodies can be used in an immunochromatographic technique to detect pLDH in whole blood (OptiMAL, Flow Inc., Portland OR).

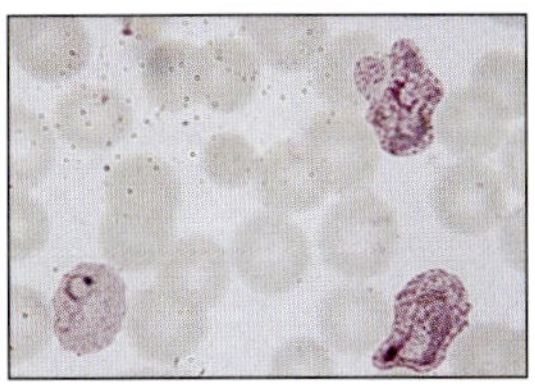

P. vivax trophozoites

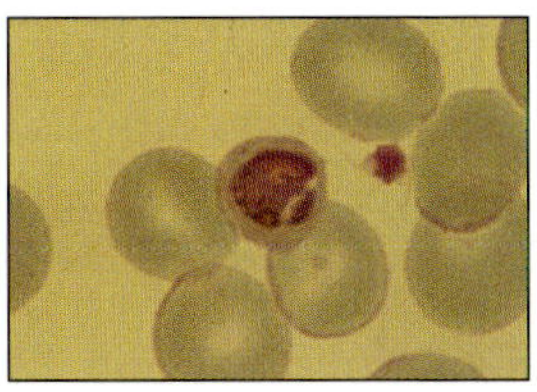

P. malariae: late trophozoite

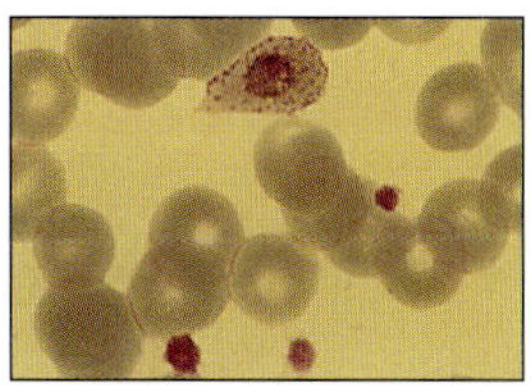

P. ovale trophozoite

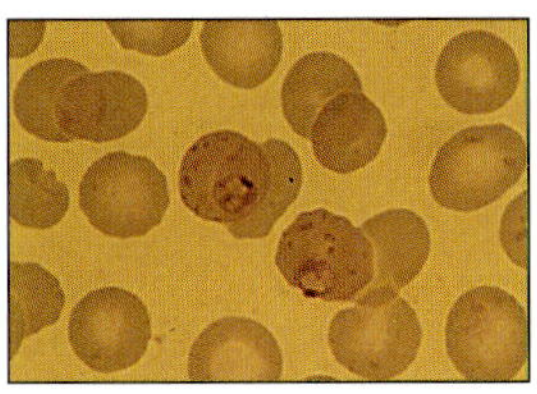

P. falciparum trophozoites

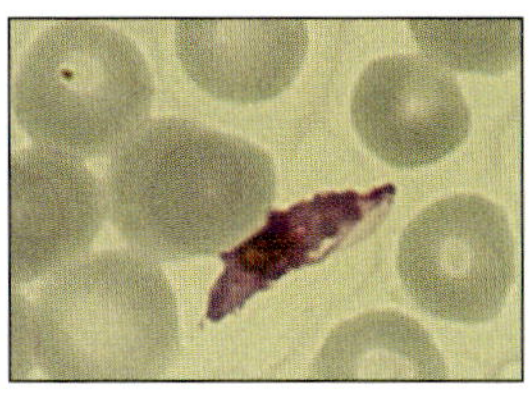

P. falciparum gametocyte

Body-fluid and tissue flagellates

Classification

Phylum — Euglenozoa

Class — Kinetoplastidea

Live in bloodstream and tissues
Vector (blood-sucking invertebrates) required

Trypanosomatidae
Single flagellum

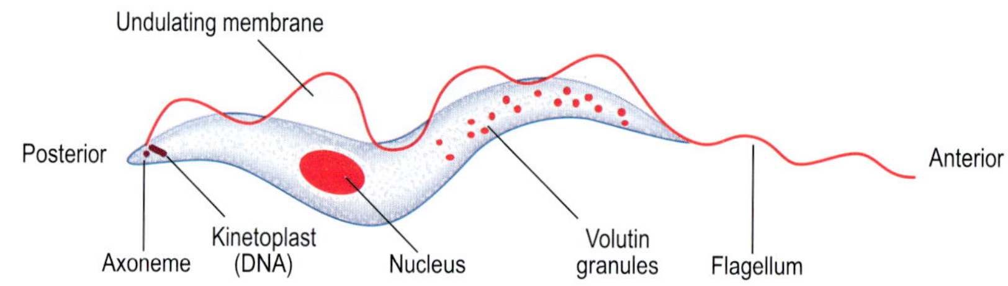

Morphological stages of the Trypanosomatidae affecting humans

***Leishmania* spp.**

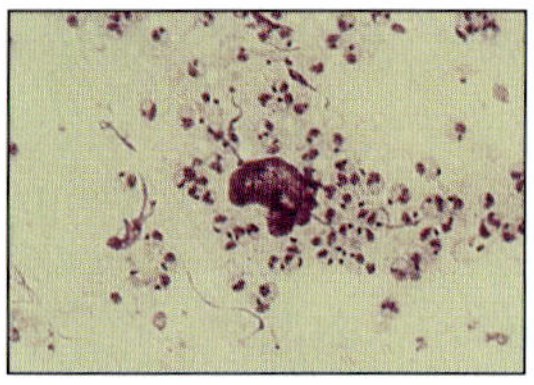

Leishmania amastigotes

Amastigote (L-D body)

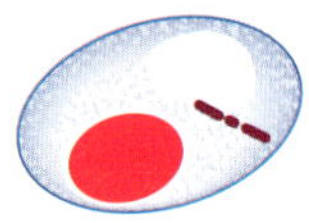

Intracellular in macrophages in humans

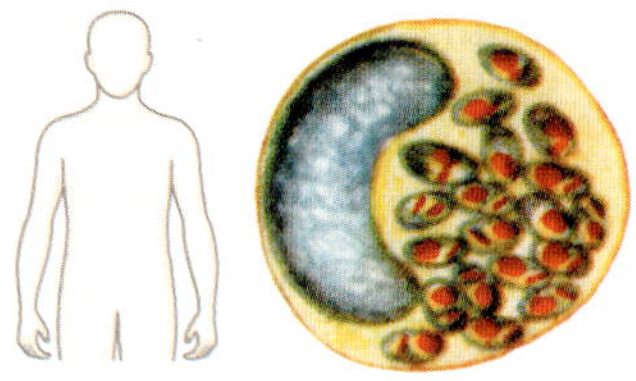

Promastigote

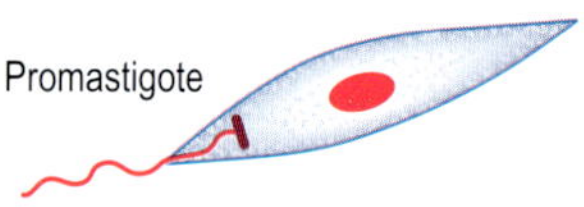

In midgut, then proboscis of sandfly (transfer stage to human, also in culture)

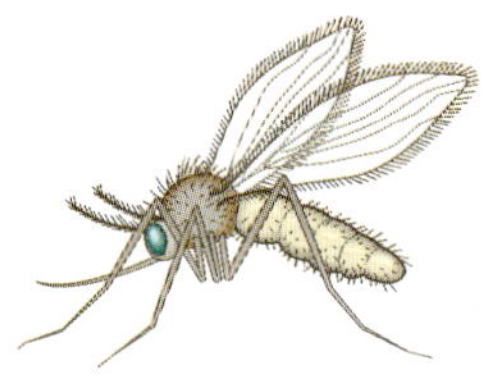

***Trypanosoma* spp.**

Trypanosoma brucei rhodesiense
Trypanosoma brucei gambiense

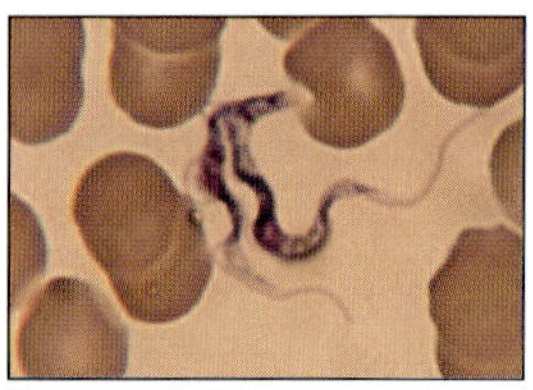

Trypomastigote

Epimastigote

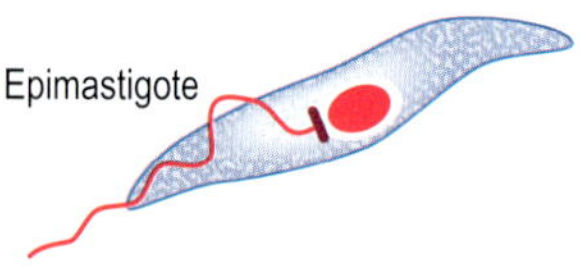

In salivary glands and proboscis of tsetse fly (transfer stage to human)

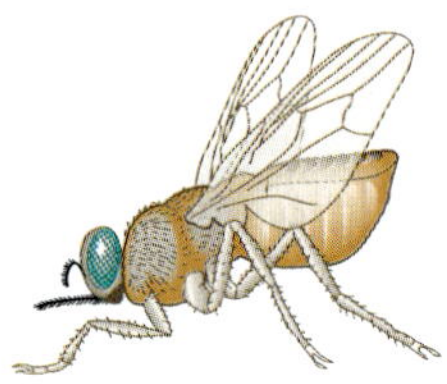

Trypomastigote

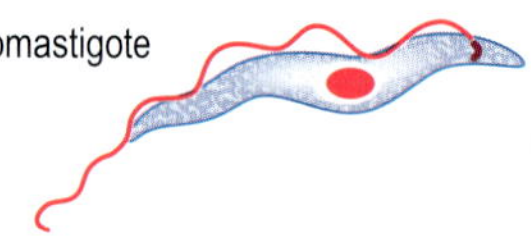

In bloodstream, lymph nodes and later CNS of humans

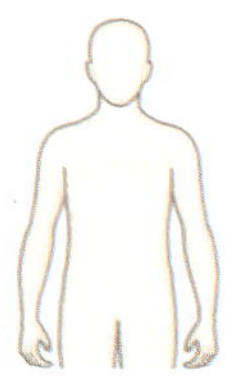

Trypanosoma cruzi

Amastigote

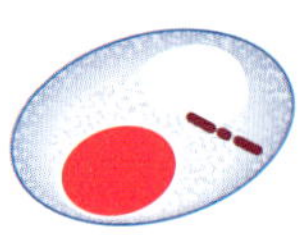

Intracellular in macrophages and tissue cells of humans

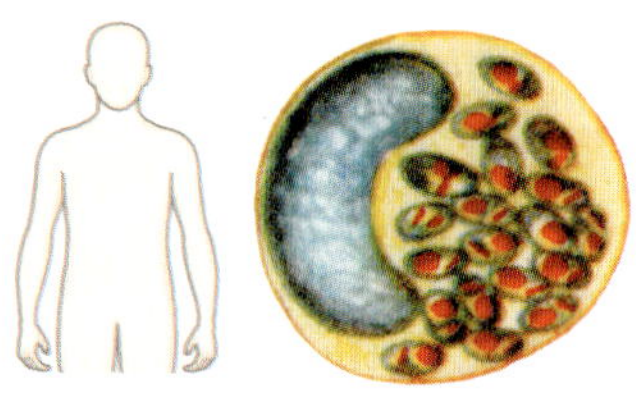

Trypomastigote

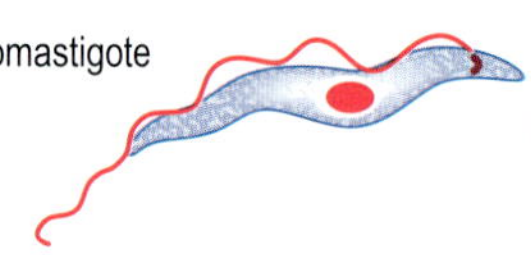

In midgut, then faeces of bug (transfer stage to humans)

In blood and tissue spaces of humans

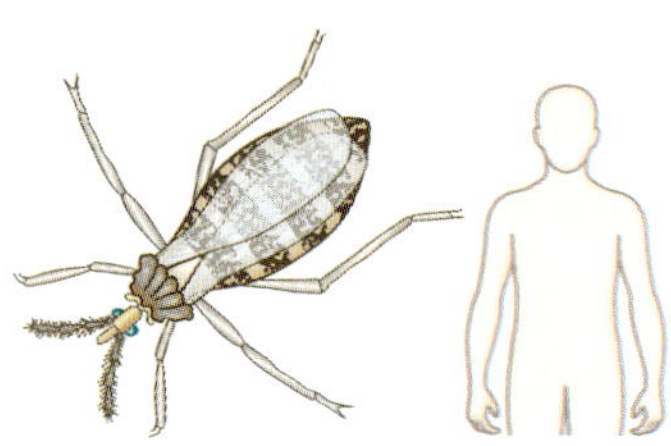

Leishmaniasis

Species	*L. donovani* complex *L. infantum* *L. donovani* *L. chagasi*	*L. tropica* *L. major* *L. aethiopica* *L. infantum*	*L. braziliensis* complex *L. amazonensis* *L. mexicana*
Disease	Visceral (kala azar)	Cutaneous	Muco-cutaneous (Espundia)

Distribution

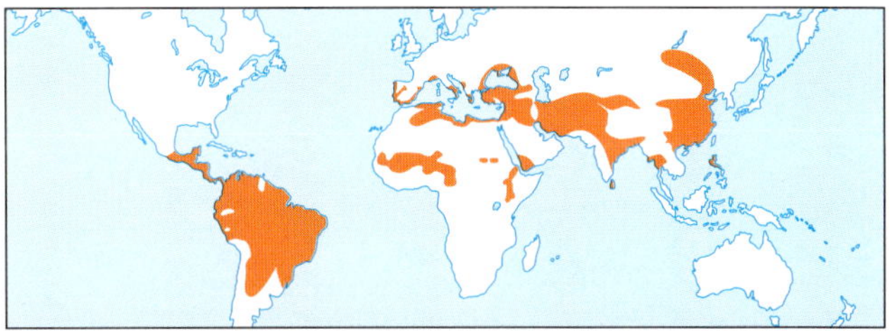

Life cycle and morphology of *Leishmania* (similar in all three species)

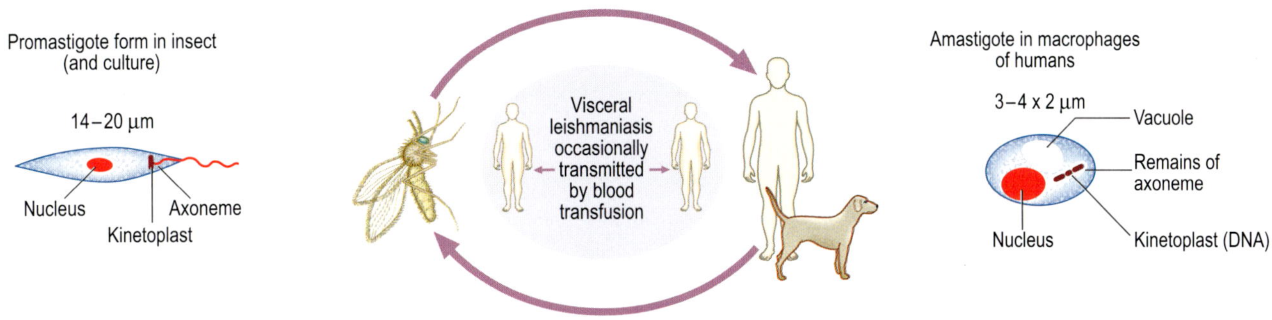

Life cycle in insect

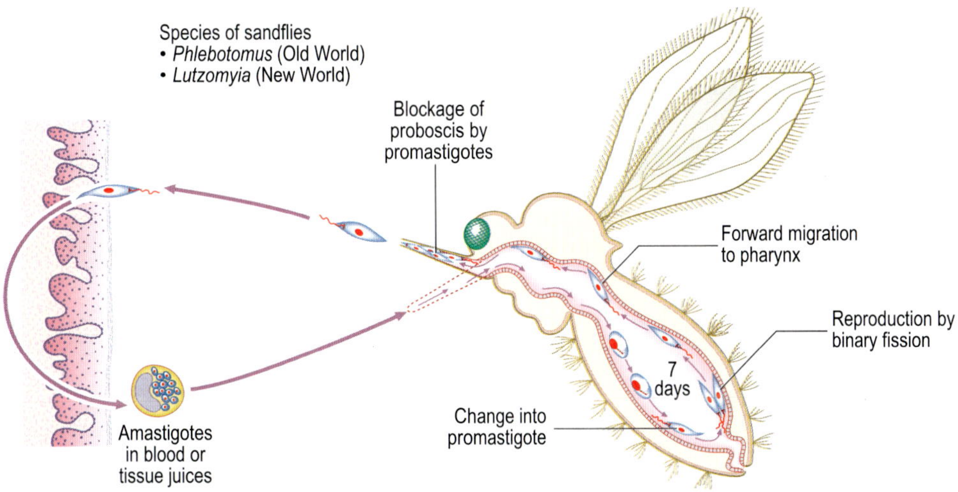

Life cycle in humans and reservoir animals

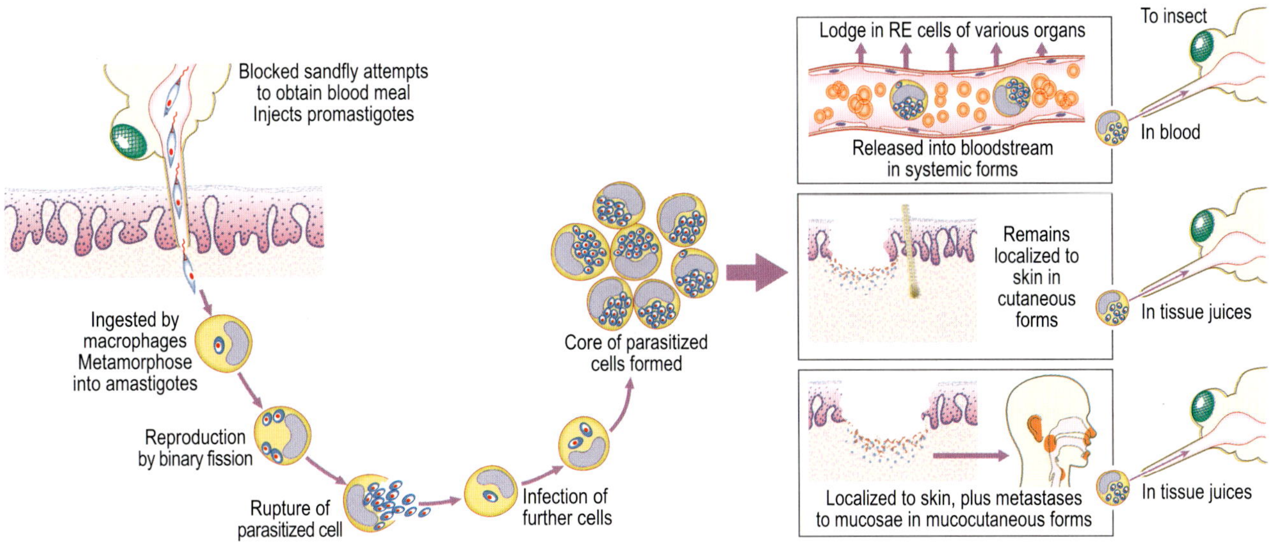

Visceral leishmaniasis (kala azar)

Distribution

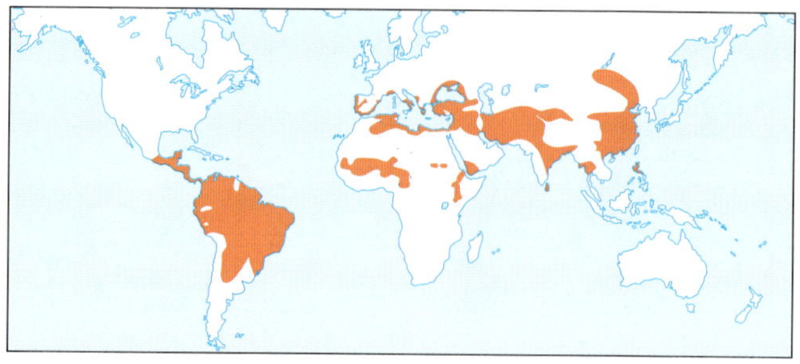

Clinico-pathological correlation

Spleen

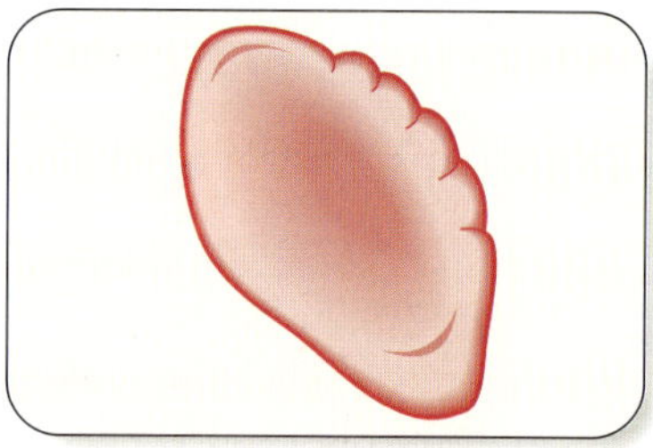

- Parasitized macrophages and endothelial cells
- Splenomegaly, pain from perisplenitis
- Spleen appears congested, dark red, soft and friable. Markedly enlarged
- The capsule is thickened and, later, infarcts and fibrosis occur

Liver

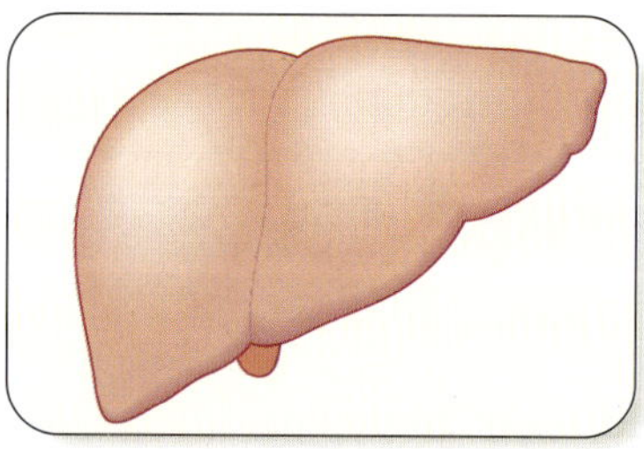

- Hepatomegaly
- Liver appears enlarged, fatty congested and later may become cirrhotic
- Parasitized proliferated Kuppfer cells with atrophy of the liver cells and later fibrosis

Lymph nodes

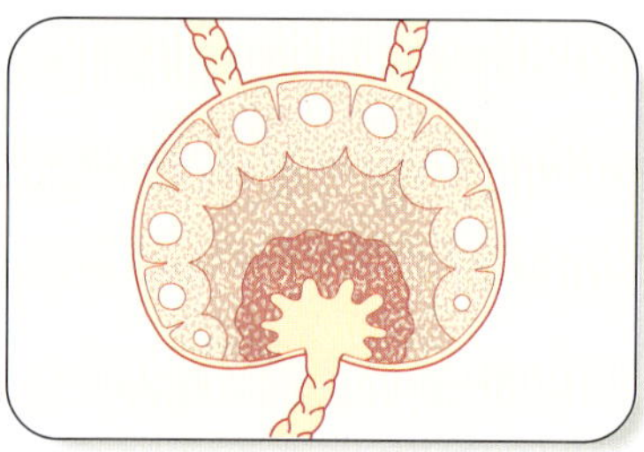

- Lymphadenopathy
- Reactive hyperplasia with parasitized macrophages

Mechanism of pathology

Offence

Skin and nasal mucosa involvement can occur (especially in Sudan) leading to ulceration → Parasitization and breakdown of RE cells. Release of parasites and debris with further parasitization → Secondary effects:

- • Displacement, degeneration of parenchymatous tissue
  • Toxaemia, cachexia fever, weight loss, weakness
  • Pallor, cardiac dilation, tachycardia, low blood pressure, haemic murmurs, ankle oedema
- • Replacement of bone marrow
  • Anaemia, leucopaenia, thrombocytopaenia
  • Degenerative myocarditis
  • Stomatitis, cancrum oris, cough, diarrhoea
  • Purpura, epistaxis, melaena
- Pigmentation: darkening of skin at forehead, temples, mouth (Kala azar means 'black fever')

Defence

Multiplication and mobilization of RE cells with stimulation of immune response. Enlargement of liver, spleen and lymph nodes → Later inconstant fibrosis-cirrhosis-haemorrhage. Jaundice, ascites, rashes → Localization in the skin after treatment Migration of parasites to skin post-kala azar dermal leishmaniasis (PKDL)

Both offensive and defensive processes give rise to increased serum globulin and reversal of a/g ratio.

Cutaneous leishmaniasis

Caused by *Leishmania tropica*, *L. major*, *L. aethiopica*, *L. infantum*, *L. braziliensis* complex

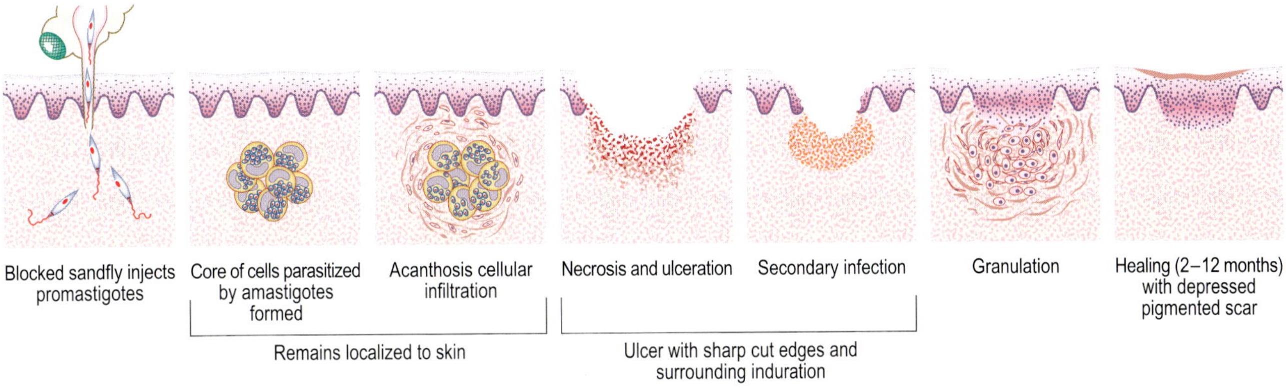

Mucocutaneous leishmaniasis (espundia)

Caused by some infections with *L. braziliensis* (Central and South America), *L. aethiopica* (Ethiopia), *L. mexicana*

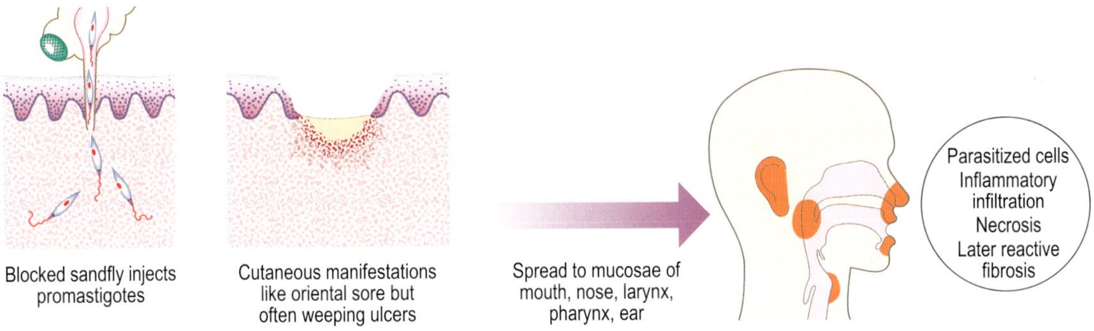

Secondary effects in loose mucosal tissues

- Oedema and capillary involvement
- Interference with local blood supply
- Necrosis - extensive destruction

- Secondary infection
  - Deep erosion locally
  - Spread of infection to lungs or elsewhere

- Healing with fibrosis

Leading to:

- Extensive disfiguring lesions
- General constitutional upset (fever, pain, anaemia)
- Bronchopneumonia and septicaemia

Diagnosis of Leishmaniasis

Visceral

Amastigotes can be demonstrated by staining bone marrow, lymph node fluid, nasal scrapings (in the Sudan), liver biopsy or splenic aspiration specimens (although this can be a dangerous procedure). Rarely, amastigotes can be demonstrated in buffy coat preparations from peripheral blood.

Cutaneous and mucocutaneous

Demonstration of the parasite is possible in stained films from slit-skin smears taken from the indurated edge of an ulcer, biopsy of the margin of the ulcer and from mucosal scrapings in mucocutaneous type.

Culture (NNN or a liquid medium such as Schneider's Drosophila medium or 199 medium with added fetal calf serum) is used for all types of material for diagnosis. Animal inoculation is rarely used now.

Polymerase chain reaction (PCR) can be used to diagnose and type the species of *Leishmania* present in biopsy or culture material.

Specific serological tests are IFAT, ELISA, direct agglutination test (DAT), or latex agglutination for IgG antibodies. An immunochromatographic test for rK39 antibody detection is also available.

African type: sleeping sickness

Caused by either *Trypanosoma gambiense* (chronic sleeping sickness, found in West Africa, the Congo, Zaire) or by *T. rhodesiense* (acute sleeping sickness, found in Zimbabwe, Tanzania, Zambia, Angola). Both have similar life cycle and morphology.

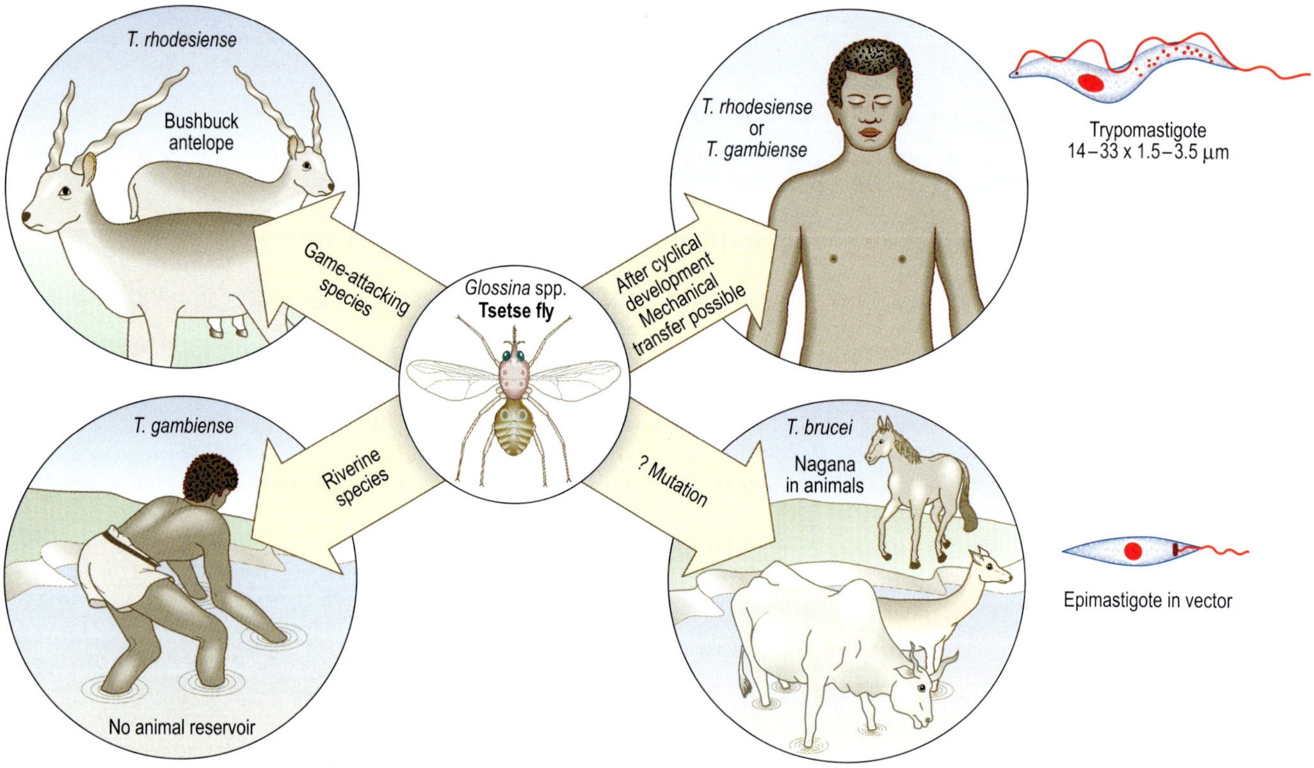

Life cycle in insect

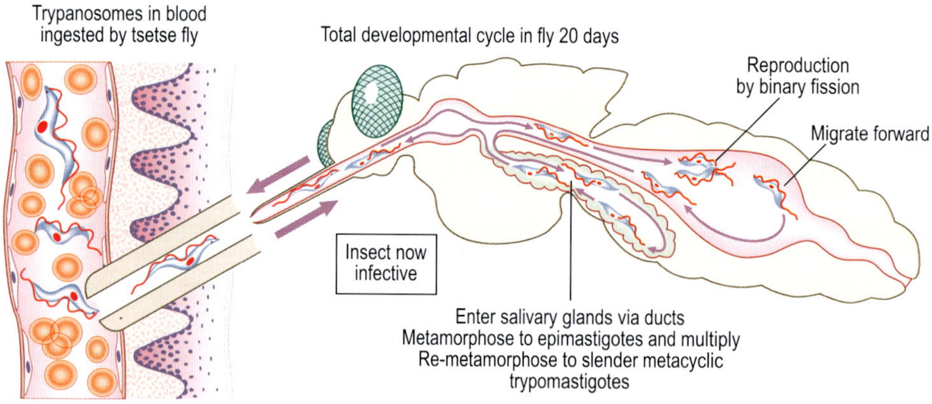

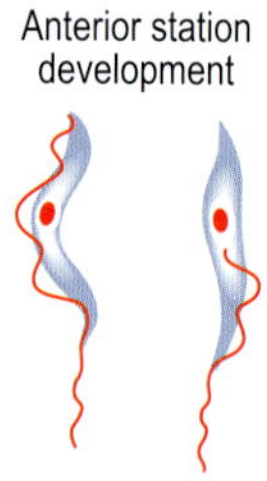

Life cycle in humans

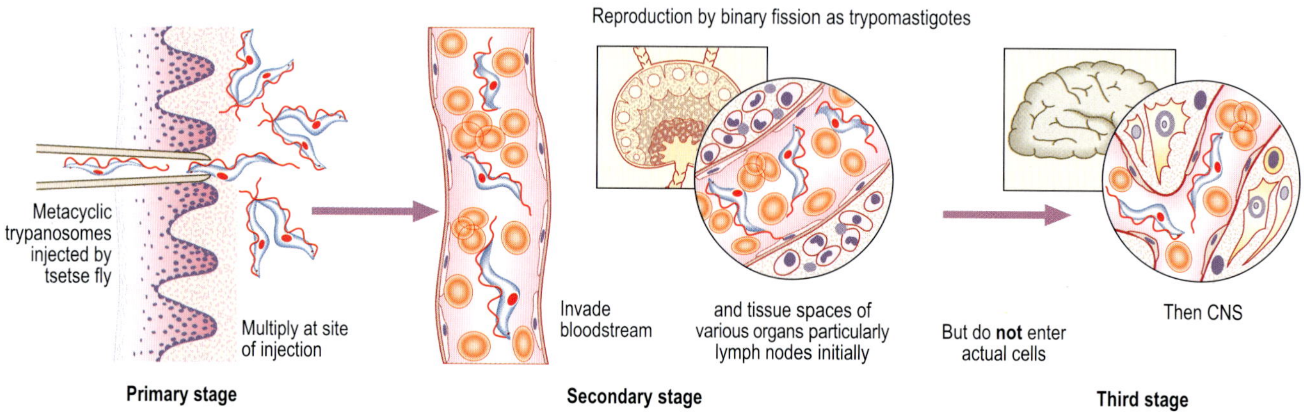

Pathogenesis and pathology

Primary stage

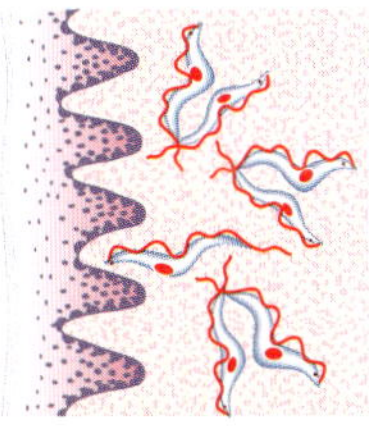

- Multiplication at site of injection
- Surrounding inflammatory reaction

Local inflammatory lesion

Secondary stage

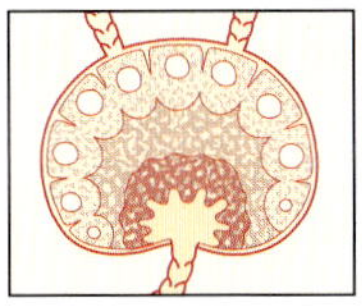

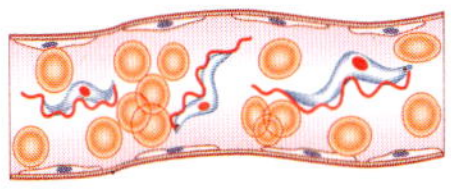

- Parasitaemia and toxaemia
- Invasion of tissue spaces (**not** cells) of various organs

Predominantly: Lymph nodes — CNS

- Damage to endothelial cells of blood vessels, surrounding (perivascular) granulomatous reactions and haemorrhages

Toxic degeneration and pressure atrophy of tissue cells

Third stage

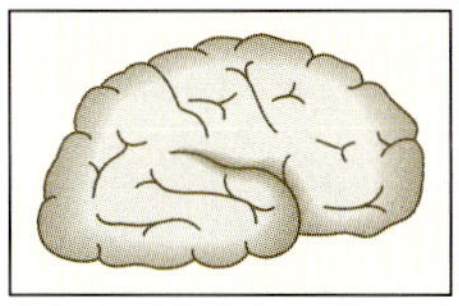

Chronic sleeping sickness (Due to *T. gambiense*) — Differ only in degree — **Acute sleeping sickness** (Due to *T. rhodesiense*)

Primary stage

Clinico-pathological correlation (chronic): Firm, tender, painful red nodule 1–3 weeks

Clinico-pathological correlation (acute): As in chronic

Secondary stage – predominantly blood and lymph node involvement

Chronic – clinico-pathological correlation

Fever
- Low
- Irregular
- Recurrent

General toxic symptoms
- Backache
- Headache
- Tachycardia
- Irregular skin rashes (circinate)
- Transient oedema face

Lymphadenopathy
- Typically post-cervical

Later anaemia monocytosis

Slight enlargement liver, spleen

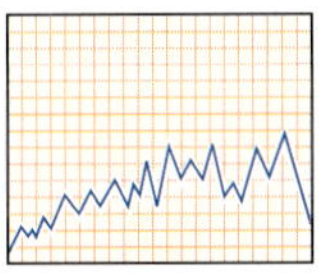

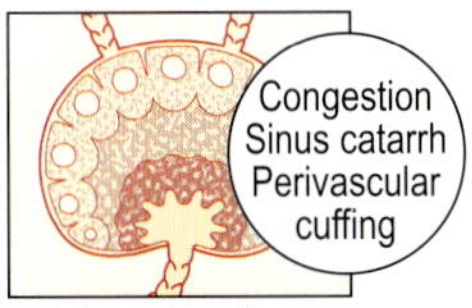

Enlarged, soft, red. Later regress, fibrotic

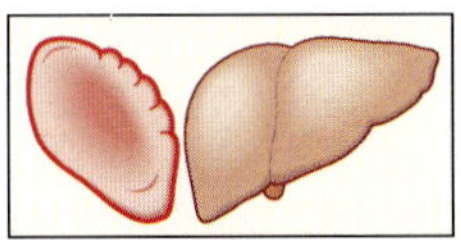

Congested, slightly enlarged

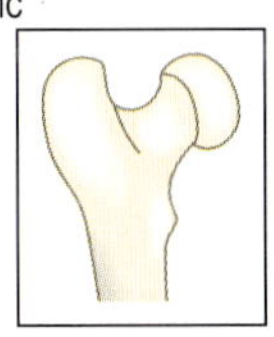

Toxic depression Bone marrow

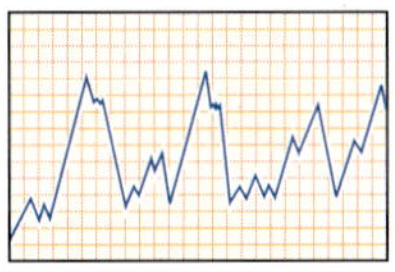

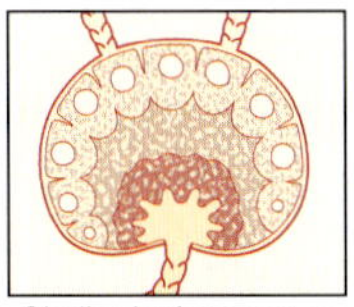

Similar lesions not so pronounced

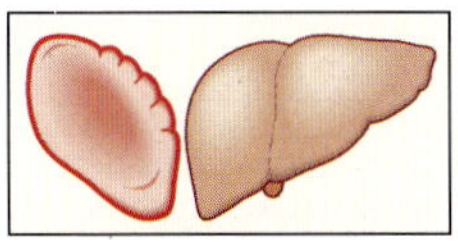

Slightly enlarged

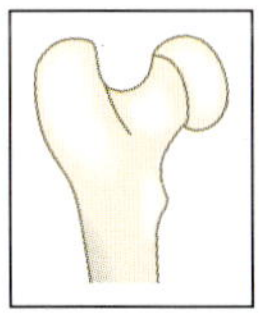

Toxic depression Bone marrow

Acute – clinico-pathological correlation

Fever
- High
- Persistent

Severe toxic symptoms
- Headache
- Vomiting
- Shivering
- Oedema face
- Serous effusion
- Bone pain

Lymphadenopathy

Myocarditis

Anaemia

Purpura

Hepatitis

Often **death** at this stage

Third stage – CNS involvement

Chronic – clinico-pathological correlation

Progressive involvement of the CNS

General symptoms of progressive encephalitis

Focal signs uncommon

Dementia

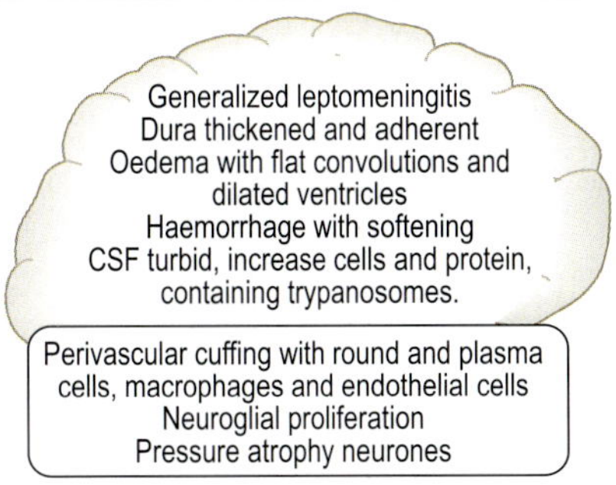

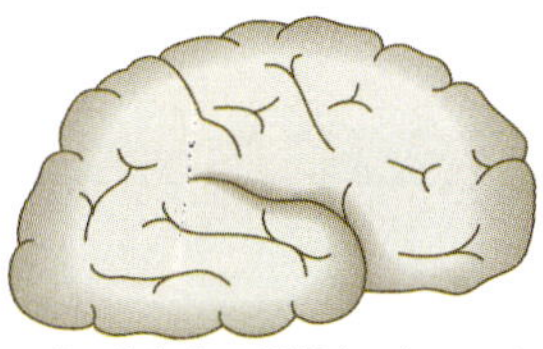

Death before CNS involvement
or
Similar changes but more acute

Acute – clinico-pathological correlation

May have early onset of encephalitis with rapid development of **coma**

Note on epidemiology – Vectors of *T. gambiense* are riverine species, hence disease often epidemic:
- *G. palpalis*
- *G. tachinoides*

Vectors of *T. rhodesiense* are game-attacking species, hence disease more often sporadic:
- *G. morsitans*
- *G. pallidipes*
- *G. swynnertoni*

South American type: Chagas' disease

Caused by *Trypanosoma cruzi*. The parasite is harboured in humans, domestic animals such as cats and dogs, and some wild animals, notably armadillos and opossums.

Morphology

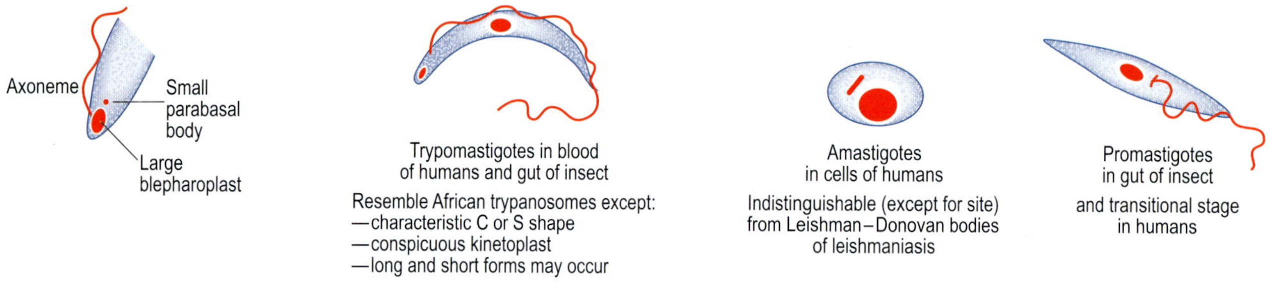

Life cycle in insect

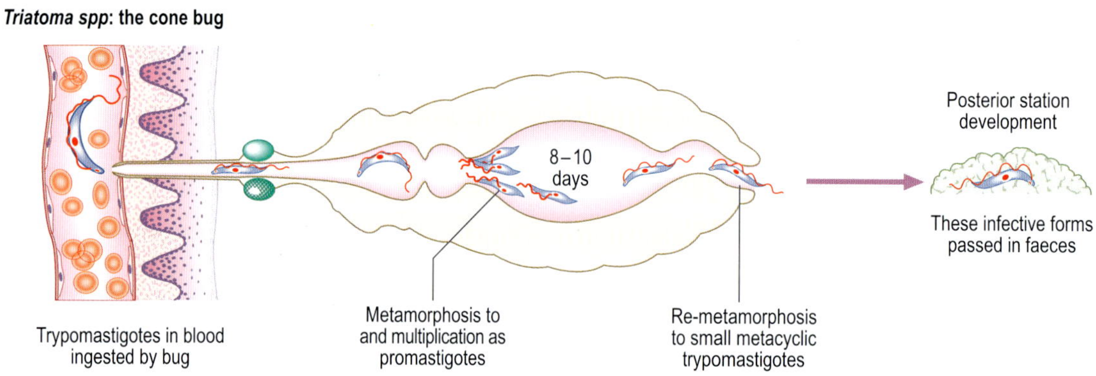

Life cycle in humans

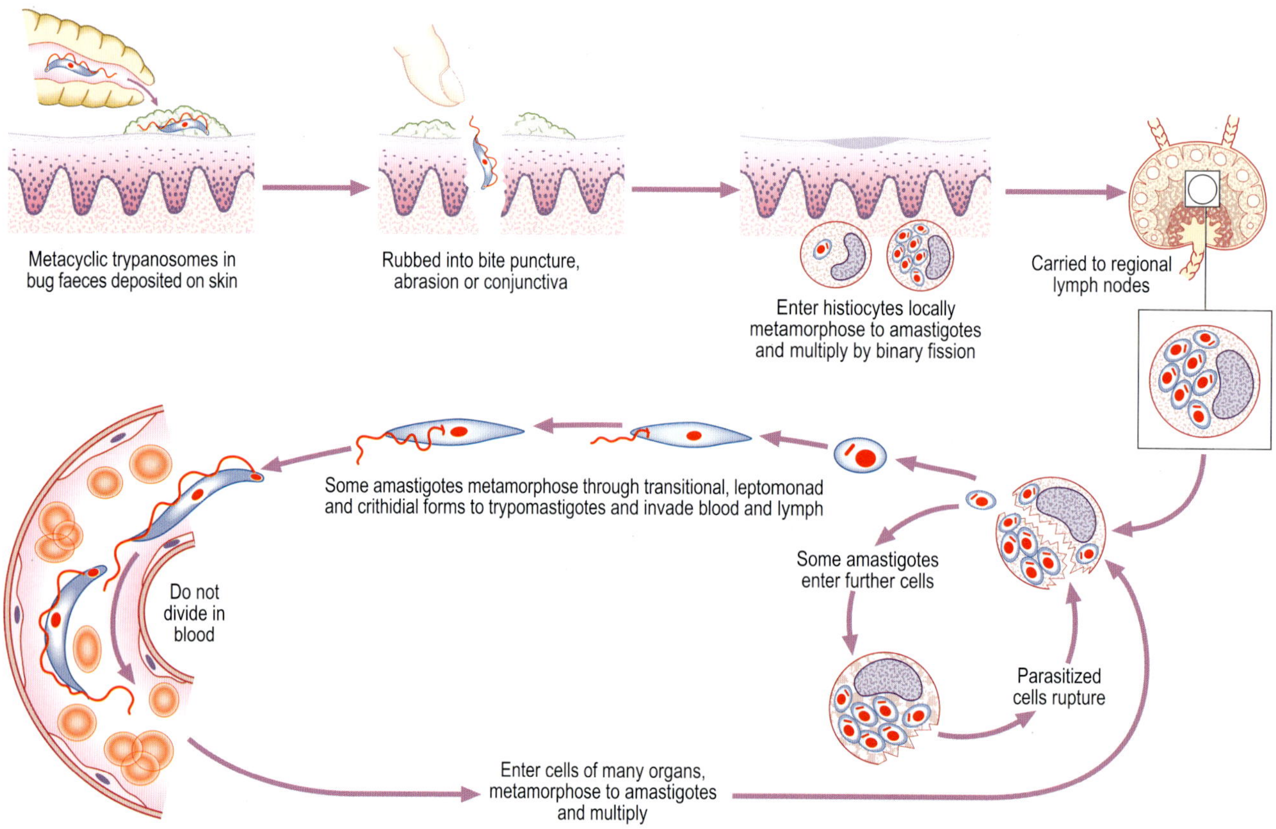

Pathogenesis and pathology

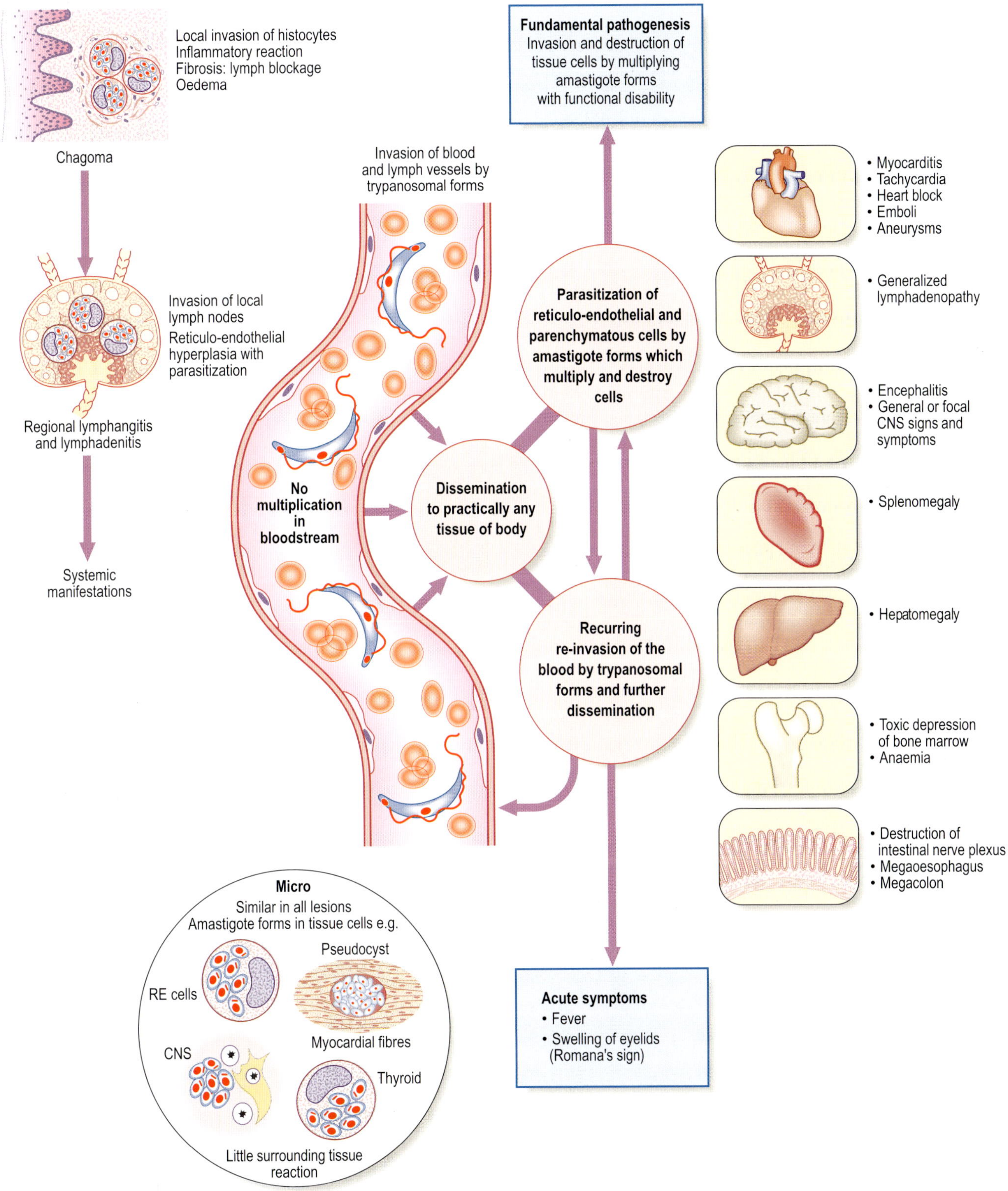

Acute form (often in children)
- Blood and reticulo-endothelial cells predominantly involved
- Fever
- Oedema
- Lymphadenopathy
- Enlargement of liver and spleen
- Sometimes encephalitis

Death or →

Chronic form
- General toxic symptoms and focal signs depending on localization
- Predominantly cardiac and CNS manifestations
- May be asymptomatic

Laboratory diagnosis of trypanosomiasis

African type (sleeping sickness)

Demonstration of the parasite.
Microscopy of thin and thick blood films and buffy coat preparations.

Trypanosomes can also be seen in smears from bone marrow and centrifuged cerebrospinal fluid (CSF). Culture is possible but difficult.

Microscopic detection of trypanosomes in peripheral blood may be improved by the use of a mini-anion- exchange column or by the use of the QBC11® (Becton Dickinson) to concentrate the parasite.

The CSF might show increased protein and lymphocytes.

South American type (Chagas' disease)

Demonstration of the parasite.
Stained smears of peripheral blood show trypanosomal forms in C or S shape. Stained films of lymph node fluid show amastigotes. It is possible to show trypomastigotes by animal inoculation from blood and by culture from lymph node fluid. Histological methods from biopsy or post-mortem material. It is also possible to demonstrate the parasite through xenodiagnosis: clean bred triatomid bugs fed on the patient's blood develop trypanosomes in the gut.

Serology (e.g. by ELISA or IFAT) is the method of choice for the detection of chronic *T. cruzi* infection.

Recapitulation

Luminal intestinal protozoa

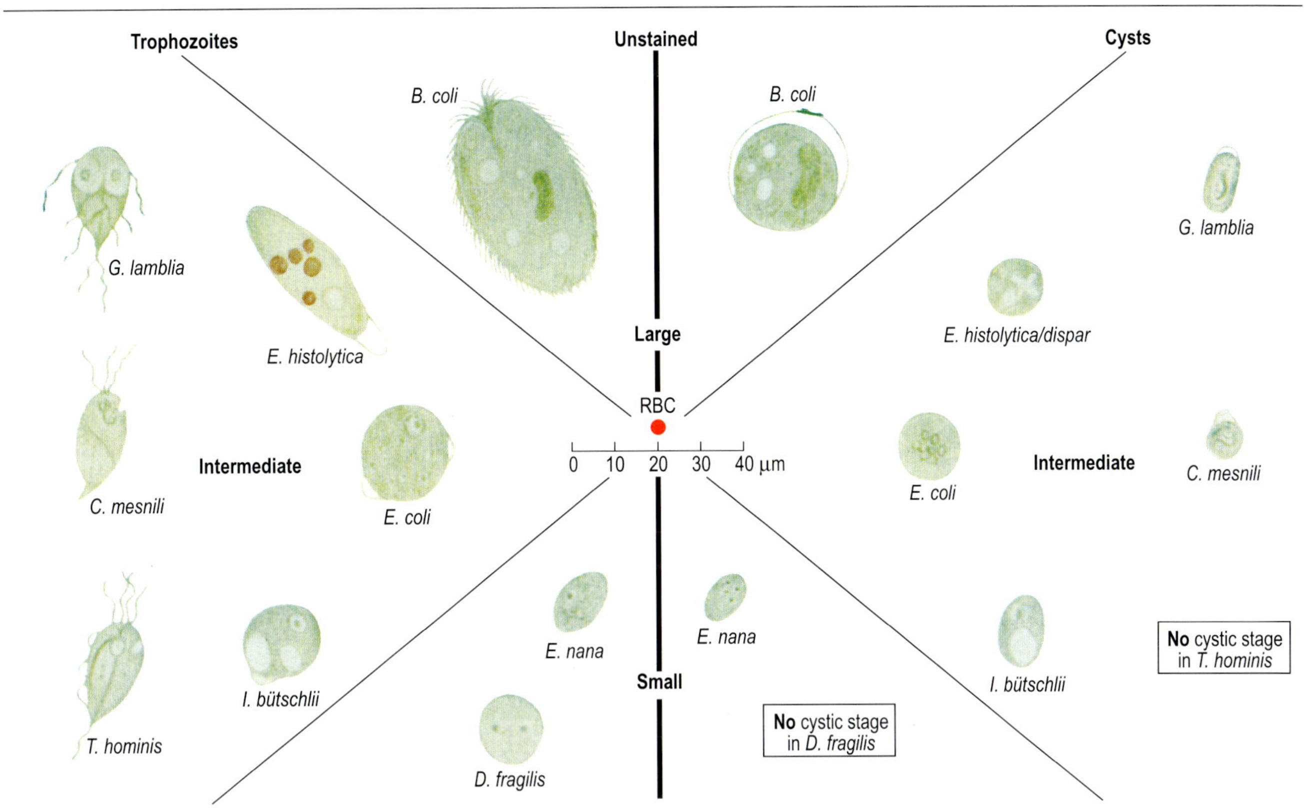

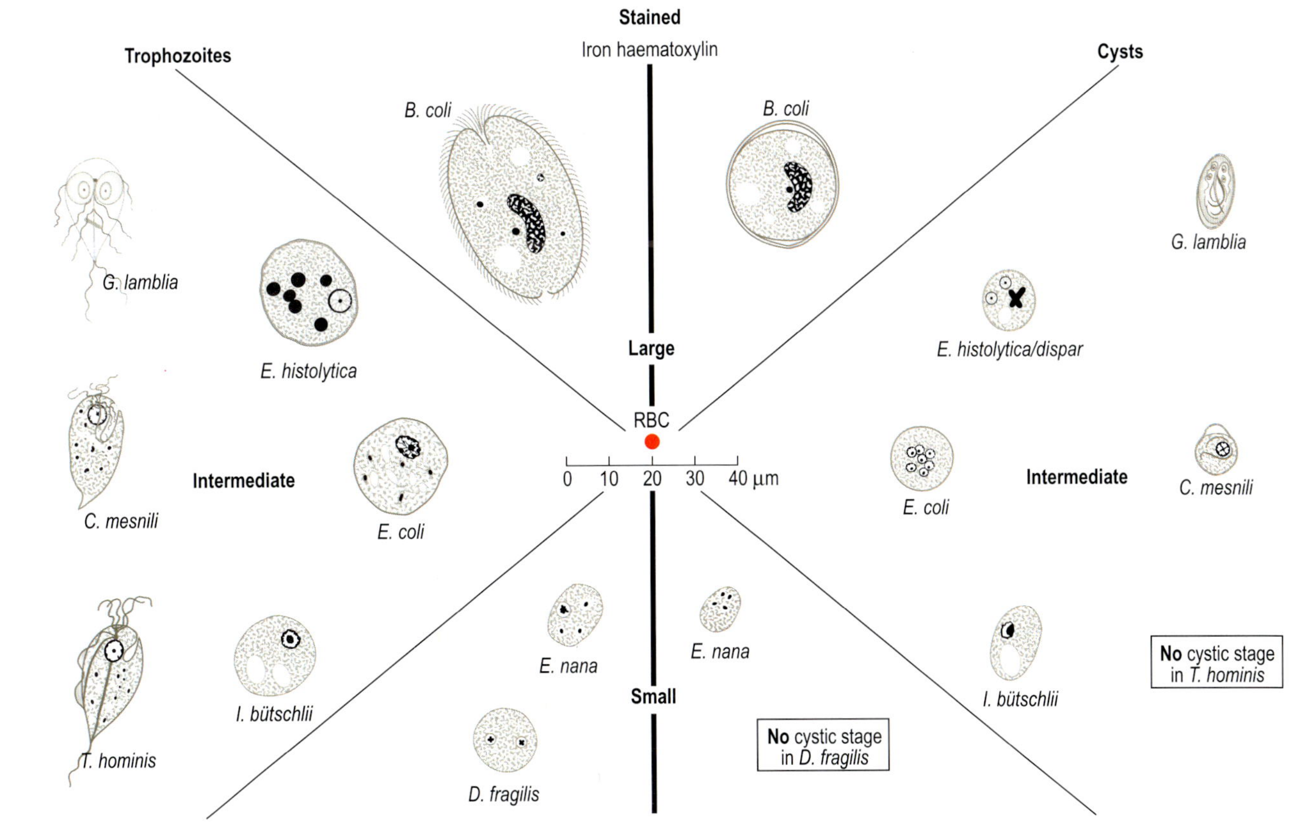

Refer to text for the following additional stained appearances:

- trichrome stain for coccidia and microsporidia spp.
- modified Ziehl-Neelsen for *Cyclospora, Isospora* and *Cryptosporidium* spp.

Index